W9-BYG-462

The Darwinian Tourist

Viewing the World through Evolutionary Eyes

CHRISTOPHER WILLS

With Photographs by the Author

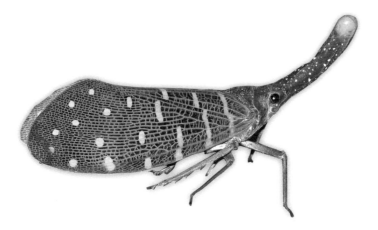

OXFORD
UNIVERSITY PRESS

OXFORD
UNIVERSITY PRESS

Great Clarendon Street, Oxford OX2 6DP

Oxford University Press is a department of the University of Oxford.
It furthers the University's objective of excellence in research, scholarship,
and education by publishing worldwide in

Oxford New York

Auckland Cape Town Dar es Salaam Hong Kong Karachi
Kuala Lumpur Madrid Melbourne Mexico City Nairobi
New Delhi Shanghai Taipei Toronto

With offices in

Argentina Austria Brazil Chile Czech Republic France Greece
Guatemala Hungary Italy Japan Poland Portugal Singapore
South Korea Switzerland Thailand Turkey Ukraine Vietnam

Oxford is a registered trade mark of Oxford University Press
in the UK and in certain other countries

Published in the United States
by Oxford University Press Inc., New York

British Library Cataloguing in Publication Data
Data available

Library of Congress Cataloging in Publication Data
Data available

Printed in China

978–0–19–958438–3

1 3 5 7 9 10 8 6 4 2

To Teodoro Francisco, who will see unimaginable wonders and perhaps greet the dawn of the twenty-second century

Why, then the world's mine oyster …
Pistol, *The Merry Wives of Windsor*

Contents

List of Illustrations

Introduction

Our understanding of the living world can always be deepened by a Darwinian perspective. As we look at the world through evolutionary eyes, we come away with a renewed sense of wonder about life's astounding present-day diversity, along with a new appreciation of that diversity's fourth dimension—its long evolutionary history. When we think about our planet from an evolutionary point of view it also becomes achingly clear that when we lose a species or an ecosystem, we lose many millions of years of history. And each such loss reduces the ecological diversity on which the survival of our species and the entire biosphere depend.

New discoveries by evolutionary biologists have a direct bearing on this loss. In recent decades it has been shown that all populations in nature are filled with genetic variability, a finding that provides yet another dimension to the Darwinian way of looking at the world. This genetic variation, like a full tank of gas in a car, provides the motive power for future evolutionary change. At the same time as we try to preserve species and entire ecosystems from extinction we must also preserve the genetic variation within species. If, through inbreeding, the few survivors of a species have lost that full tank of evolutionary gas, then even if we manage to nurse that species back to apparent health it may remain on the critical list for centuries.

The preservation of genetic variation has a direct bearing on our own survival. Many of the crop species and domesticated animals on which we depend have lost critical variability. Through our own ignorance and mismanagement we have driven many of their wild relatives to extinction, losing reservoirs of critical variation in the process.

In this book we will embark on a series of adventures that illustrate this evolutionary way of looking at the world. To help communicate the impact that these experiences have had on me I have provided photographs that I took in the heat of the moment.

We will begin with a look at the teeming underwater life of Indonesia's Lembeh Strait, and then explore how this diversity arose from distant simple

Figure 1 The Nilgiri Hills in southern Tamil Nadu state nurture forests at their base, providing protection for some of India's most endangered ecosystems. Their wet uplands also give refuge to some of India's oldest human tribes. The hills serve as a bridge between the history of our species and the history of the natural world.

ancestors. We will go to a little valley in northern Israel, to probe the factors that drive evolution—mutation, natural selection, chance events, and genetic recombination—and see how these processes bring about the emergence of new species.

We will encounter an earthquake that recently hammered me while diving on a coral reef off the island of Yap, and expand our view of that local event to take in the multitude of catastrophes large and small that have shaped the fragile crust of the planet. And we will explore how such geological events have shaped evolution in unexpected ways.

Immersing ourselves in the teeming life of rainforests, we will see how the process of evolution has maintained complex ecosystems over time, and how evolutionary forces have caused rainforests species to become more diverse over time. And we will show how these same forces have driven us apart from our own nearest relatives, the apes, making us dramatically different.

We will visit the dry valleys of western Mongolia, where long ago humans encountered wolves and began to tame the natural world through artificial selection. As we follow the history of domestication, we will see how, again and again, our ancestors hunted the wild relatives of domestic animals to extinction and closed off vast evolutionary possibilities in the process.

We will walk in the footsteps of the first people to leave Africa on the greatest migration our species has undertaken, and see how they were forced to adapt to the sweltering jungles of Southeast Asia and the searing deserts of Australia. And we will follow close relatives of ours who had made that same migration earlier, only to be eventually doomed to extinction even though they shared many traits with us. If, as seems likely, our ancestors helped to drive these people to extinction, how many possible evolutionary paths were closed off in the process?

Everywhere we travel, we will see how evolution and ecology have interacted to yield the world we live in. This interaction will be especially clear in our visits to the tropics.

When we think of a tropical forest we visualize a wet, dense, and tangled riot of plants, plunged into near darkness at ground level by giant trees buttressed overhead. But the tropical forest of southern India's Mudumalai National Park does not conform to this image. It is sunny, dry, and open, characterized by relatively small native teak, myrtle, and almond trees. The precipitous Nilgiri Hills to the east provide a consistent source of groundwater that percolates into the forest, but rainfall is uncertain and the dry season lasts for half the year.

As a result the forest is often swept by fire, which keeps the undergrowth down. The wild and semi-wild elephants that abound at Mudumalai are able to crash cheerfully about, stripping branches from the trees and sometimes smashing the tree trunks into fragments.

H. S. Dattaraja (Datta to his many friends) and his colleagues at Bangalore's Indian Institute of Sciences have spent two decades following the changes that have taken place in the forest. They have repeatedly carried out painstaking censuses of all the trees in a half-square-kilometer plot, so that they can follow what happens to this piece of forest over time.[1] I have played a small role in this project, by providing them with some statistical tools that help to explain why Mudumalai and other tropical forests remain so diverse.

The results obtained by this group of ecologists show that evolutionary and ecological processes have maintained the diversity of this forest. Their research gives us a glimpse of a past world even more packed with plant and animal species than the present, a world that we will explore in Chapter 5. But their data have also shown how, as human influence grows stronger with each passing year, this ecosystem is beginning to lose its unique character. As Datta and I tramped through the forest, we found that it had been invaded by alien plants that we humans have spread throughout the world—the ubiquitous *Impatiens* and *Lantana* flowers, and the tough and resourceful prickly pear cactus. As we change the world's ecosystems, we are bringing new evolutionary pressures to bear on the plants and animals that live in them.

I have spent most of my biological career in the laboratory, applying deliberate artificial selection to produce evolutionary changes in simple organisms with brief lifespans. Then, about fifteen years ago, I joined a collaborative project to look at the diversity of Mudumalai and other tropical forests, which make up some of the world's most complex ecosystems. It has been most rewarding to analyze the dynamic changes taking place over decades in these forests. And it has been equally rewarding to find that the diversity of these ecosystems is maintained by processes similar to those that evolutionary biologists study in the laboratory. Working with this world-wide group of collaborators, I was able to show that the evolutionary force of natural selection acts in tandem with ecological processes to increase the diversity of the world's ecosystems.

In the course of this work I have had the pleasure of traveling around the world, both on the land and beneath the surface of the oceans. Some of these

Figure 2 (*opposite*) H. S. Dattaraja surveys the Tropical Forest Dynamics Plot at Mudumalai in southern India that has inspired much of his scientific career.

Figure 3 Lantern flies of the genus *Phocia*, at Lambir in Sarawak, Borneo, one of the forest plots that we are studying in order to investigate the evolution of the immense biodiversity of tropical ecosystems.

trips have been in conjunction with the rainforest project and other research projects. I must admit that some have simply been for fun. But regardless of where or why I traveled to the world's wild and inaccessible places, I tried to view them through evolutionary eyes.

Every biologist knows there is much more going on in the living world than we can possibly duplicate in the simplified, sterile confines of the laboratory. Over the past two decades I have tried to share my excitement about this larger biological world in a series of books about evolutionary subjects. Those books explored how and why genes can sometimes become faster at evolving, how we can use data from the Human Genome Project to examine our own evolution, how we have evolved as a species and may continue to evolve in the future, and how life itself might have begun. All these themes contribute to this book, but the primary lesson that emerges from these far-flung tales is how much we can enrich our experiences by bringing an evolutionary perspective to our travels.

This book's publication coincides with the International Year for Biodiversity, a multifaceted program of research and education about the world's ecosystems that is coordinated by the United Nations. It also coincides with an unprecedented explosion of worldwide ecotourism. Never before have so many people, including many of the readers of this book, visited so many distant and exotic destinations. I hope to share with you my own delight at being a Darwinian tourist, and to encourage you to view your own travels through a Darwinian perspective.

Many people contributed help, advice, and logistical support to this book. I especially acknowledge my editor, Latha Menon, my indefatigable travel agents Barbara Jenkins-Lee and Jenny Collister, and my many guides around the world, especially Tugso Tugso in Mongolia, Roland Ranaivo-Ratsitohaina in Madagascar, Dougie Wright in Botswana, John Wapimaro in Tanzania, Lay Maung in Myanmar, Mike Kuiper and Diane Corbiere on Yap, Sonia Goggel in the Solomon Islands, Kay Samsen in the far reaches of the Andaman Sea, and Cedric Lesenechal beneath the waters of the Spice Islands. I have relied on colleagues in many different disciplines for help and advice: Ofer Bar-Yosef, Liza Comita, Richard Condit, H. S. Dattaraja, Charles Davis, Scott Edwards, Pascal Gagneux, Nimal and Savitri Gunatilleke, Pam Hall, Kyle Harms, Hopi Hoekstra, Steve Hubbell, Robert John, Andy Knoll, Abraham Korol, John Lawrence, Albert Lin, Suzanne Loo de Lao, Elizabeth Losos, Richard Meadow, James Moore, Mike Morwood, Piotr Nasrecki, Cheryl Nath, Supardi Noor, Miguel Rivera, Oliver Ryder, I-Fang Sun, Sylvester Tan, Jill Thompson, Matt Tocheri, Anne-Marie Wills, and David Woodruff. Special thanks go to Mark Strickland, who helped me with many aspects of underwater photography. And of course many thanks to my wife, Liz, who has badgered me across six continents and around innumerable islands, making sure that I don't leave things behind.

Despite all this help there are undoubtedly sins of both omission and commission in this book. The responsibility for all of them falls on my own balding head.

A note on the photographs

I used Canon D60, 20D, and 1DIII digital cameras, and most of the pictures started as raw files. There are a few scanned photos from the ancient days of film. Underwater pictures were taken with the aid of a marvelous and faithful Ikelite DSLR housing and two Ikelite strobes. Many of the pictures have been cropped, and I employed temperature, tonal curve, and contrast adjustments to improve the color and tonal balance. This was especially necessary for the underwater photographs, in which even the best-lit subjects often appear too blue. As is common in photography I often used dodging and burning to emphasize the subject in the picture relative to its background. And in the underwater pictures I used Photoshop to remove those annoying little creatures that float about in the water and reflect the light back to the lens in the form of little bright spots. I have not, however, removed, added, or rearranged any other objects in the pictures.

PART I
The Living World

The giraffe-necked weevil, *Trachelophorus giraffa*, of Madagascar uses its long neck like a crane to roll leaves and build a nest. In this part of the book we will look at how such astonishing adaptations evolved, in the context of the vast changes that have taken place on our planet during geological time.

1

Shape-Shifters

Figure 4 A spectacular Ceratosomid nudibranch crawls along the dark bottom of Indonesia's Lembeh Strait. At Lembeh, as perhaps nowhere else on Earth, we can explore the full range of our far-flung distant relatives on the evolutionary tree.

As I hovered above a sandy bottom, twenty meters down in the Lembeh Strait, I surprised a Pharaoh cuttlefish that was gently snoozing above a small coral outcrop. Its outline was broken up by the many small lumps and protuberances all over its body, and its tentacles were crumpled into irregular shapes like complex pieces of origami. A blotched color pattern of green and yellow completed its disguise as an apparent lump of coral.

When I moved closer it woke, swiveled alertly to face me, and began to change its appearance. The many protuberances melted away in seconds, so that its body became smooth. Its arms and tentacles unfurled and straightened. Its color became lighter and more uniform. Within a minute it was transformed from an almost invisible coral-like object into a streamlined animal ready to flee. As I took my last picture of this whole-body makeover, the cuttlefish jetted away with a pulse of water from its siphon and disappeared into the darkness. The cuttlefish is one of the world's most accomplished shape-shifters, and it had just given me an effortless demonstration of its skill.

Surely no two organisms could be more dissimilar than the ingenious and graceful water-breathing cuttlefish and its clumsy air-gulping human observer. But in fact, even though present-day cuttlefish are expert shape-shifters and we are not, we had a common ancestor. And, at the time of that common ancestor, a far more astonishing shape-shift took place, one that had enormous evolutionary consequences.

How do we know that we are related to the cuttlefish? When and how did we first take such different evolutionary paths, and how have we and the cuttlefish converged in some of our abilities? What other animals have branched off from our different lineages during the long course of our evolutionary divergence? And is it possible to investigate, and perhaps even to recreate, the events that took place at the distant time when we and the cuttlefish began to diverge?

As a good Darwinian tourist, these evolutionary thoughts spun through my mind as I watched my remote relative propel itself into the dark.

Figure 5 A Pharaoh cuttlefish, *Sepia* sp., caught sleeping. Its body is covered with retractable protuberances and its tentacles are crumpled, so that it resembles a lump of coral.

Our immense family tree

When Darwin briefly visited the geologically young Galàpagos Islands in 1835, he was overwhelmed by evidence that recent evolutionary changes had shaped life on that remote archipelago. It gradually became clear to him that the closely related animals and plants on the different islands of the archipelago had radiated adaptively from a small number of ancestors that had made their way or been carried to the islands. His visit to the Galàpagos, along with many other observations that he made during his five-year voyage, helped to plant the germ of the idea of natural selection in his mind.

When I plunged into Indonesia's Lembeh Strait my experience was very different from Darwin's. I was overwhelmed by the almost insane diversity of life there. Traces of recent evolution are not common at Lembeh, though

Figure 6 The cuttlefish, newly streamlined, is now ready for its getaway.

there are some fish and other animals unique to this narrow passage between islands. But these examples of recent evolution are almost lost in the cacophony of more than half a billion years of evolutionary divergence.

The Lembeh Strait lies at the heart of one of the world's great biodiversity hot spots, where there is a greater variety of marine life than anywhere else in the world.[1] It is one of the waterways that surround the world's eleventh largest island, a jewel of rich tropical diversity called Sulawesi that lies just to the east of Borneo.

Sulawesi has been so thoroughly pushed and twisted by tectonic forces that the map of the island looks like a character from some forgotten alphabet. Indeed, the Portuguese explorers who first landed on different parts of Sulawesi's complex and deeply indented coastline were fooled into thinking that it was actually several islands. They named this hypothetical archipelago the Celebes, perhaps a mishearing of Sulawesi, which in turn may be derived from local words meaning "iron island."

Figure 7 An effigy of the deceased leads the coffin of its owner to its final resting place in Tana Toraja, Sulawesi.

The island is almost as large as Great Britain, but its biology is far richer. On land its ecosystems range from rainforest to grassland, encompassing a wide variety of animal and plant life. We will meet some of these remarkable terrestrial organisms later.

It was this exotic but still easily accessible world that I left behind when I took a diver's giant stride and splashed into the waters of the Lembeh Strait, which separates the tiny island of Lembeh from the main island's northeast coast. Above the surface the strait is a rather undistinguished narrow channel of water, flanked on each side by forested hills. The waters of the passage are deep and sheltered enough to make it safe for coastal shipping, though it can be dangerous for ocean-going vessels. The strait's shores are blemished by undistinguished lumbering and fishing towns, but beneath the surface its biological diversity has made it a mecca for scientists and scuba divers from around the world.

Coastal vessels have used the strait as a shortcut for centuries, and their sailors have tossed empty bottles and other trash over the side. The trash settles to the sandy bottom, where it is soon partially buried.

When I first entered the water at Lembeh, the thought of all that junk waiting for me on the bottom was less than thrilling. Lembeh's underwater world is far from glamorous. Fabulous coral gardens adorn other parts of Sulawesi's coast, but there are no extensive coral reefs here. Reefs cannot become established because the strait is repeatedly scoured by strong oceanic and tidal currents of nutrient-rich water. As a result corals grow only in small patches, wherever there is something solid that they can use as an anchor.

Instead of a maze of colorful corals I was greeted by a level plain of dark sand and mud that stretched off in all directions, broken only by islands of eel grass and a few coral-covered outcrops. By stretching out horizontally and using minimal fin movement to avoid stirring up the mud, I was able to swim smoothly from one clump of coral- and weed-covered detritus to the next.

Most marine animals live, not in the open water, but in what is called the benthic zone. The benthic zone is defined as the ocean bottom and the space immediately above it, along with a maze of burrows and secret places that lie just below the surface. Although the word benthic comes from the Greek benthos, meaning the deep sea, even shallow waters have benthic zones.

Organisms that inhabit benthic zones battle endlessly for space to live, with an intensity that would put Southern California real estate developers to shame. In Lembeh these battles ensure that each clump of overgrown debris on the bottom is covered with a riot of intensely competing creatures.

The fish of Lembeh provide a logical place to start to explore our immense family tree. Unlike most of the creatures that live on the bottom of the strait, fish are vertebrate animals that are quite close to us in evolutionary terms, such that we can all feel an immediate kinship with them. And yet even these close relatives of ours have evolved in unexpected directions.

Among the shyest of these diverse fish are the pygmy seahorses, a mere centimeter long or less, that make themselves seem even smaller by curling their tails around the branches of pink and orange sea fans. The sea fans dine on tiny free-swimming arthropod plankton that they snare using their stinging cells. Because the seahorses are unable to trap the plankton themselves, they browse delicately along the branches of the sea fans, nibbling on the tiny creatures trapped there.

These little equine fish show an uncanny resemblance to the branches of the fans on which they live. The colors of their bodies and the little warts on their skins help them mimic the details of the surfaces of the sea fan branches with precision.

I found one of these seahorses, whitish with pink bumps, clinging in a strong current to an actively feeding sea fan. Its pouch was swollen with hundreds of tiny young, so it was clearly a male. The pregnancies of male seahorses and pipefish provide one of the clearest cases in the natural world in which the roles of the sexes are reversed.[2] The seahorse female, after giving up her eggs to the male, has blithely left her progeny behind and moved on to sexual pastures new. During her reproductive life she will compete fiercely with other females and attempt to mate with as many other males as possible, each of whom will serve as incubators for her offspring.

Seahorses are only a small sample of the fish diversity of Lembeh. Consider the frogfish, which come in dramatic shades of red, white, pink, black, and green. They nestle on the bottom, in the branches of corals, and among the fronds of algae. Members of a single frogfish species can adopt different colors, depending on where they are trying to hide. They only reveal themselves when they open their mouths to suck in innocent fish. Frogfish have such capacious mouths that they have occasionally been seen to eat other frogfish almost as big as themselves.

In a clump of detritus lurked a spiny devilfish, which lived up to its name—with its glaring eyes and its tooth-filled, downturned mouth it is the stuff of nightmares. Its wing-like pectoral fins and the spines on its back are covered with weeds and other growths that effectively conceal its outline. It has evolved a surprisingly insect-like mode of locomotion, crawling forward on appendages formed from parts of its pectoral fins.

Figure 8 (*opposite*) A pregnant male pygmy seahorse, *Hippocampus bargobanti*, at Lembeh Strait. The males nurse the babies while the females are free to seek other mates. This seahorse was clinging to a fan coral in strong current, and in the background you can see the actively feeding polyps of the coral.

Figure 9 A painted frogfish, *Antennarius pictus*, lies in wait for its prey, which can include other frogfish.

The sandy flats between the different islands of detritus swarmed with life too. Gurnards and sea moths, shaped like stealth bombers, stirred up the bottom as they rowed across it using their fanlike pectoral fins. Goggle-eyed balloonfish cruised by, their spines ready to deploy whenever they swelled with water to scare off enemies. Black-and-white-striped convict snake eels writhed swiftly across the bottom in search of prey. In their shape and color, these eels mimic the air-breathing and highly poisonous sea snakes. One of the convict eels thrust its head swiftly into the sand right in front of me, moving so quickly that I could not see what tiny unfortunate animal it had caught.

Even objects that must surely be dead turned out to be alive. Many brown dead leaves from the nearby forests fall into the strait and litter the bottom. They drift along in the current at odd angles, as dead leaves would be expected to do. But on close examination some of these leaves turn out to be brown scorpionfish—leaf-shaped, leaf-colored, and covered with

Figure 10 This spiny devilfish, *Inimicus didactylus*, grows protuberances on its back that soon become covered with algae and other small creatures. It crawls along the bottom on leg-like modified fins.

vein-like patterns and irregular splotches that make them look even more like a real leaf.

The real leaves that drift along the bottom have been part of the scenery for millennia, and the scorpionfish have evolved not just to look like them but even to mimic how they drift. Like their brightly colored frogfish relatives, the scorpionfish wait until their incautious fish prey swim too close, under the blithe misapprehension that there is surely nothing to fear from a dead leaf.

More distant branches on the Lembeh family tree

We humans have a relatively close evolutionary kinship with all these fish, even with the rather creepy spiny devilfish. Despite our decidedly different shapes and habits we all share a backbone, and this shared trait places us all in the subphylum Vertebrata. But if we venture a little further among the

Figure 11   A hunting convict snake eel, *Elapsopis versicolor*, writhes swiftly across the bottom. These snake eels imitate the air-breathing and highly poisonous sea snakes.

spreading branches of our family tree we find other slightly more distant relatives. Some of these, unlikely as it may seem, are sea urchins.[3]

Jostling crowds of large sea urchins, known as fire urchins, are common at Lembeh. They form dense clusters, swarming with surprising speed across the sand and sucking up small creatures from the bottom as they go. Their spines, some long and striped and others purple-black, radiate out in all directions to protect their plump (and delicious) bodies.

As I peered down at this carpet of spines I immediately discovered why these roistering ragamuffins are called fire urchins. Their bodies, glimpsed among the spines, are colored the fiery red of hot coals. The red patches are outlined in electric blue spots that glow like sparks.

How do we know that these sea urchins share an evolutionary kinship with scuba divers and merchant bankers? At the end of the nineteenth century the English zoologist Walter Garstang compared the early embryonic stages of vertebrates with the early stages of sea urchins, starfish, and other echinoderms. He found that vertebrates and echinoderms have similar early development, and that this shared development differs markedly from the early embryonic stages of other large groups of animals such as insects and mollusks.[4] We may like to think that we have more in common with a hardworking and loyal honeybee than with a tousled and uncharismatic sea urchin, but the evidence of our shared youthful anatomies says otherwise.

It is not surprising that right down until the 1960s Garstang's conclusion was rejected by some other anatomists. This is in part because he went too far, and concluded that vertebrates had sprung from ancient echinoderm stock. We now know that our common ancestor probably didn't look much like either modern echinoderms or modern vertebrates. But it is now clear that he was right about his essential point, that we are indeed closely related to the echinoderms. A century after Garstang's pioneering studies, the subtle signs of kinship that he drew from the anatomy of early development were reinforced by molecular studies. Comparisons between echinoderm and vertebrate DNA sequences prove our close relationship beyond a doubt.

Figure 12 A leaf scorpionfish, *Taeniotus tricanthus*, drifts along the bottom, doing a most convincing imitation of a dead leaf while waiting for nearby fish to be fooled.

Figure 13 Fire urchin, *Astropyga radiata*. These urchins are close relatives of ours, even though they do not seem to resemble us in the least.

Several other great branches of the tree of life, more remote from us than the echinoderms, are represented in abundance on the seafloor of the Lembeh Strait. One of them is the great phylum Arthropoda, made up of the animals with jointed legs.

Members of this phylum swarm everywhere on the bottom at Lembeh. Decorator crabs lurch out of their hiding places like camouflaged tanks whenever a diver poses a threat. Their swollen legs, covered with carefully groomed mini-gardens of sponges and small algae, look like the limbs of some body builder who has gone overboard on doses of growth hormone.

Figure 14 An almost invisible commensal anemone shrimp defends its sea anemone home.

Nearby are anemone shrimp, which defend their homes on pink and purple sea anemones. It is easy to see the surfaces of the anemones right through these shrimp, for they seem to be made of oddly shaped bits of glass. The only clues to the presence of these delicate transparent arthropods are bright medallions of color that accent their otherwise imperceptible bodies. If it weren't for their blobs of opaque color and their tiny striped eyes that float on the ends of transparent stalks, they would be invisible.

Other shrimp are colored and patterned all over. The marbled shrimp are the last word in understated elegance. One of them slowly inches forward out of its lair. Its designer-patterned carapace gives it a decidedly upscale air compared with the roistering crowd of less fashion-conscious creatures that surround it, though the effect is somewhat diluted by the shaggy moustache of cirri that it uses to ensnare its food.

The local king of the arthropods is undoubtedly a red, green, and blue shrimp, one of several species of brightly-colored mantis or boxer shrimps

Figure 15 An elegant marbled shrimp of the genus *Saron* uses brushlike cirri to snare its prey.

at Lembeh. It peered at me from its burrow in the sand through round robotic eyes on stalks. Its thick claws, ending in bulges shaped like a box-er's glove, were held at angles in front of its body like the claws of a praying mantis.

By any name these shrimp are formidable customers, even though they are only about twenty centimeters long. Their claws can snap forward at 80 kilometers an hour, moving so quickly that cavitation bubbles form in their wake.[5] As the bubbles collapse they actually generate flashes of light.

The claws are quite strong enough to crack a diver's face mask. Mantis shrimp have even been known to blast their way out of glass aquarium tanks. One assumes that the escapees enjoy a fleeting moment of triumphant free-dom before they expire on the aquarium floor.

My shrimp scuttled swiftly out of its lair to defend itself. I chose discre-tion over valor and moved my vulnerable camera and facemask well back from its claws, leaving the field to this tiny action figure.

Figure 16 This mantis or boxer shrimp, *Odontodactylus scyllarus*, patrols the sea floor. The shrimp's claws can move so quickly through the water that cavitation bubbles form in their wake. As the bubbles collapse they generate flashes of light.

The incredible mollusks

The mollusks are the most diverse of this otherworldly zoo of creatures. We are of course familiar with the clams, oysters, mussels, and squid featured in all the world's great cuisines. These happen to be the mollusks that have committed the evolutionary mistake of being delicious. But the mollusks are far more diverse than restaurant menus would imply, and many mollusks are not tasty at all. More importantly from an evolutionary viewpoint, many of the mollusks have capabilities and lifestyles far beyond our reach—or at least beyond the reach of everybody but a shape-shifting comic book superhero.

The mollusks, like the arthropods, are distant from us on the animal evolutionary tree. But there is no doubt about our ultimate kinship with them, because we share divergent but still detectably similar DNA sequences.

Like the arthropods, the mollusks occupy their own phylum. The family tree of the mollusks can be followed back through the fossil record for more than half a billion years, chiefly because the limestone shells that many of them construct make excellent fossils.

The name mollusk simply means "soft of body," a catchall category if ever there was one. Pioneering Swedish taxonomist Carolus Linnaeus grouped them into the single phylum Mollusca, recognizing the anatomical similarities among their diversity.

Many mollusks are protected by hard shells, and live most of their lives in one place. Here at Lembeh, giant Tridacna clams are firmly buried in the mud amid clumps of coral. These clams have shells so large that they have often been used for baptismal fonts in churches. These huge animals must filter vast amounts of water for food. They get an additional shot of energy from tiny photosynthetic symbiotic algae that live in their soft mantles and give them their richly patterned blue, green, or brown colors. When the clams reproduce they spew forth great fountains of eggs or sperm into the surrounding water. The resulting larvae are carried away in the (mostly vain) hope that they will find somewhere to settle and eventually grow up into new giant clams like their parents.

Tridacna clams stare blurrily at the world through thousands of tiny window-like eyespots that can perceive only light or dark. Other shelled mollusks have more elaborate eyes that are highly sensitive to motion. Some of these eyes are faceted like the eyes of insects. Others are like miniature versions of the reflecting telescopes used by astronomers. All of these eyes, however, only form impressionistic images of the animals' surroundings.[6]

Other members of this great phylum have abandoned sight but embraced movement. The shell-less snails called nudibranchs ("naked gills") are the dazzling butterflies of the sea. They come in a rich variety of colors and patterns. On the seafloor at Lembeh I encountered an especially gorgeous nudibranch, the giant pink-colored *Ceratosoma trilobatum*, as it slid down a clump

Figure 17 This colorful nudibranch, *Ceratosoma trilobatum*, advertises to predators that it tastes terrible.

of coral and algae. A full ten centimeters long, it displayed exuberant clusters of spots that made it look like a glass of pink champagne.

Why are the nudibranchs so colorful? Like the colors of the fire urchins their bright colors have obviously not evolved for sexual attraction, since the nudibranchs cannot see each other. They can only sense the presence of potential mates chemically, by using a pair of feathery olfactory organs called rhinophores. (You can see the rhinophores at the nudibranch's head near the bottom of Figure 17. At its rear, near the top of the picture, a cluster of naked gills adorns the nudibranch's body like a flower.)

It is clear from many experiments that the colors and patterns of the nudibranchs are actually sending warnings to other species that might otherwise be tempted to eat them. These sea snails store toxins collected from the small organisms that they eat, making them highly poisonous. Like the showy colors and patterns of some butterflies, the colors of the nudibranchs have evolved as a signal to predators that they are nasty-tasting and dangerous.[7]

Of course, this warning coloration will only succeed if the nudibranchs' predators exhibit some degree of sophistication. The animals that prey on the nudibranchs must see them clearly enough to detect their warning colors and patterns, and must also be smart enough to be able to recall previous unpleasant encounters with similar nudibranchs. It is likely that in much earlier times, when the world was patrolled by more stupid and forgetful predators with poorer vision, animals like the nudibranchs that depended on warning coloration would not have evolved. The evolution of nudibranchs in their full glory depended on the emergence of smarter predators from that early stupidworld.[8]

Other groups of mollusks have moved far beyond the clams and nudibranchs in sophistication. They have been able to harness both sight and movement to aid in their hunt for prey and for mates.

On a wide stretch of nearby sea bottom a tiny cuttlefish scoots along. It raises its stumpy arms cautiously as it pauses on the sandy plain.

Like octopuses, cuttlefish have eight grasping arms, but they also have two longer, extremely prehensile tentacles. Even schools of swift silvery fish are not safe from a cuttlefish as it hunts. Such fish tend to be invisible to most

Figure 18 A flamboyant cuttlefish, *Metasepia pfefferi*, with eyes that produce sharper images than human eyes.

predators because they take on the color of the surrounding water when seen from below. But cuttlefish can find them, because their eyes are sensitive to the polarized light that the fish reflect.

The animal in Figure 18 is known as the flamboyant cuttlefish, and its name is deserved—it blazes in ever-shifting patterns of purple, yellow, and red as it scans the surface of the bottom for small prey. Its eyes are shielded by curtain-like membranes that make them appear hooded and sleepy-looking, but in fact they are keener than our own.

Because of evolutionary convergence, the eyes of the flamboyant cuttle-fish closely resembles ours. Their eyes, like ours, are camera-like marvels of evolutionary engineering, with pupils that let in light, irises that can dilate or contract to control the amount of light that enters, and crystal-clear lenses that refract the light and form an image on the retina. Like the lenses of our eyes, their lenses can change shape to focus the images of near and far objects. This is a big improvement over the awkward cameras that we humans carry around, with their clumsy focusing rings and motors that move the rigid glass lenses back and forth.

Our eyes and the eyes of the cuttlefish can both be traced back to a simple eyespot possessed by our common ancestor. This ancestral eye consisted of a few photosensitive cells, perhaps overlaid by a layer of transparent cells that protected them and concentrated the light. The evolutionary path that eventually led to the fancy capabilities of cuttlefish eyes diverged from the one that led to the equally fancy capabilities of our own eyes.

The cuttlefish eyes are better than ours in at least one important respect. They can form crisp images of their surroundings in full color across the entire span of their light-sensitive retinas. We have to be content with a fuzzy image that has a little clear spot in the center.

Why are the cuttlefish eyes better? The difference can be traced to how our eyes and those of cuttlefish develop. During embryogenesis our light-sensitive retina begins as a hollow ball of cells called an optic vesicle at the end of a stalk of brain tissue. The back region of this ball differentiates into pigmented light-sensitive cells and the front region becomes the nerve cells that will pick up the retinal signals. The two regions then collapse and fuse into a single cup-like structure. The result is a retina in which the nerve cells lie on top of the retinal cells. The nerve cells interfere with image formation, which is why most of our vision is blurred. The only clear part of our visual field is the fovea, a small region in which the nerve cells fan away from the underlying retinal cells so that they do not interfere with the image. You are looking at these words through your fovea.

The cuttlefish eye develops differently. An optic vesicle develops from its brain as well, but the ball does not dimple inwards and form two layers of cells. Instead it forms a single layer of nerve cells. Meanwhile, part of the outer layer of the embryo's developing head moves in to bond with this layer. It is this piece of ectoderm, rather than the optic vesicle tissue, that differentiates into the light-sensitive pigment cells. The result is a retina that gets it right. The nerves that transmit visual signals to the brain form a layer behind the retinal tissue rather than in front of it, so that they do not interfere with the image.

As I swam slowly closer to the flamboyant cuttlefish, our gazes met briefly across an immense evolutionary divide. I formed an image of the cuttlefish at the same time as it formed a much sharper image of me. We regarded each other through eye lenses that, like our retinas, were also constructed through different developmental pathways even though they have converged on similar structures. Starting with the simple eyes of our remote ancestor, the eyes of cuttlefish and humans have diverged and then converged again to provide this moment of mutual regard.

On another part of the flats I glimpsed for a moment the most talented of the local mollusks, the mimic octopus. This octopus, fast-moving and swift-burrowing, is a master of disguise, the Scarlet Pimpernel of the underwater world. As it flashed past me and disappeared, its arms were striped black and white, like a writhing collection of convict eels or sea snakes. Depending on the threat that it must defend against, this octopus can turn itself into a passable imitation of a lionfish, a sting ray, or a mantis shrimp. It can also imitate the movements of these dangerous predators. If all else fails, this molluskan changeling can transform itself into a clump of innocuous-looking brown seaweed.⁹

The mimic octopus accomplishes these feats by changing the color, pattern, and surface texture of its arms and body, just as the cuttlefish we met at the beginning of this chapter was able to transform itself into a good imitation of a clump of coral.

As I swam in a haze of delighted astonishment around the sandy bottom at Lembeh, I realized that it is an ideal place to explore the animal family tree. Most scuba divers in the tropics explore coral reefs, not gray muddy sea bottoms. These divers swim among a range of creatures as diverse as those at Lembeh, but many of them may remain invisible because there are so many places to hide in a coral reef. Here on the volcanic sand bottom of Lembeh, in the space of a dozen dives, I was able to contemplate a sampling of the full range of native exotic creatures, often observing and photographing their behaviors for minutes at a time.

Diving at Lembeh and at similar areas in Indonesia and Papua New Guinea is a relatively new activity, dating back only to the 1980s. Australian divers were the first to venture into these apparently unpromising shallow waters, at a site called Dinah's Beach in eastern Papua New Guinea. They were as dazzled by their experience as I was by Lembeh. They called their adventure "muck diving," after the mud and detritus that characterizes such sites. The name, if you will forgive me, has stuck.

Because so much of the evolution of life has taken place in the oceans, it is not surprising that life's diversity confronts us more vividly below the waves than above them. In a walk through the rainforest that shrouds the mountains near the Lembeh strait you will see a wide variety of flowering plants and many vertebrate animals. You will also encounter many insects and other arthropods such as giant centipedes. But if you swim around Lembeh's sandy bottom you will immediately find large and colorful animals from a far wider collection of phyla, ranging from the most primitive sponges to the most complex mollusks and vertebrates. The sea is an evolutionary time machine.

Muck-diving through evolutionary time

What would we have seen if we had been able to go muck-diving in earlier times? If we could travel back far enough, we would find a very different world.

The Solar System, including the Earth, formed four and a half billion years ago, but we can only trace the history of the Earth's crust back four billion years. Before that time our newly formed planet was being pounded mercilessly by massive objects from outer space, so that its crust was melting and solidifying repeatedly. After the crust finally cooled and was stable enough to accumulate oceans and support life, living organisms appeared surprisingly quickly, probably around three and a half billion years ago. Yet, for the first three billion of those years, any time-traveling muck-divers that ventured from the lifeless land into the oceans would have swum over a seemingly dull and uniform sea bottom. The divers would have seen featureless mud flats dotted with layered concretions of bacteria and algae called stromatolites.

This apparent dull uniformity masked a great deal of evolutionary activity that gave rise to some of life's most essential capabilities. Many different kinds of bacteria evolved soon after the first appearance of life. Some of them were able to photosynthesize, and some lineages of these bacteria eventually bequeathed these abilities to multicelled organisms that became the higher plants. As a result of these new ways of manipulating the environment to extract energy, the very chemical composition of the atmosphere and the oceans gradually changed, making the world's environment more like that of the present time.

Very little of this activity has been preserved in the fossil record. A few traces of possible bacteria may have been found in Australian rocks as old as three and a half billion years, although the exact nature of these early fossils is embroiled in controversy. Stromatolites were plentiful throughout the early history of life, but the oldest ones do not show clear signs of being built by layers of bacteria and may have simply been the result of geological processes.[10]

About 625 million years ago this superficially rather boring world of living organisms began to change. A scattering of modest-sized and extremely odd creatures with no obvious affinity to present-day organisms began to leave traces in the fossil record. These mysterious creatures make up the Ediacaran biota, named after regions in Australia where they were first found.

Even this collection of creatures, exciting though they were in comparison to the dull bacterial communities of earlier times, would have seemed pretty dull to our time-traveling scuba divers.[11] Although frond-like structures dotted the sea bottom like waving feathers, and strange flat creatures slithered among them, most of the Ediacaran organisms, like those that preceded them, were still too small to be seen with the naked eye. And yet, as we will see, this simple world might have provided an environment for evolutionary experimentation that would not have been possible during either earlier or later times.

Then, 542 million years ago, at the start of a geological period called the Cambrian, everything changed. Starting with a burst of small shelled mollusks, a multiplicity of animals soon appeared, presaging a world more like our own. The start of the Cambrian was like the beginning of a concert after an unconscionably long period during which the orchestra seems to have been merely tuning up. The sudden commencement of this full-throated evolutionary concert was so dramatic that geologists have named it the Cambrian explosion.

We have a good idea of what the bottom of early Cambrian seas might have looked like. Shale beds from the Chengjiang area that lies to the south of Kunming in southern China are filled with a wide variety of beautifully preserved fossils, prevented from decay by sudden underwater landslides. They have been dated to 525 million years ago, a mere 17 million years after the start of the Cambrian. Thriving communities of arthropods, mollusks, worms, chordates, and many other animals covered the bottom. Muck-divers in those shallow seas would have been entertained by this great diversity of creatures, though because of the lack of smart predators they would probably not have been as colorful as the creatures of Lembeh today.

The Cambrian explosion and the roots of animal divergence

The fossil record appears at first blush to show that the diversity of animal phyla arose rapidly at the start of the Cambrian. But the diversification of these animals began well before the Cambrian. At Chengjiang it is already clear that our chordate ancestors and the early mollusks were taking different paths.

A little chordate-like creature called *Cathaymyrus* from Chengjiang is the earliest animal with affinity to ourselves that has yet been found anywhere. But even this early hemichordate was already the proud possessor of gills, a heart, and a dorsal nerve chord.

The underwater landslides at Chengjiang preserved clusters of *Cathaymyrus*. These little animals apparently burrowed together in groups in the mud, like a present-day primitive hemichordate called *Amphioxus* that they resembled.

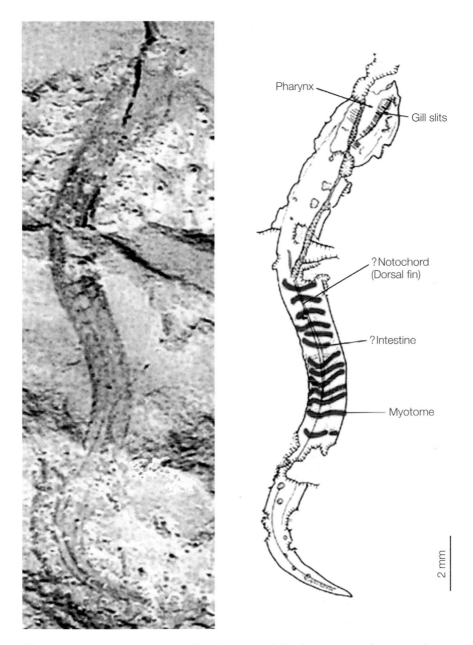

Pharynx

Gill slits

?Notochord
(Dorsal fin)

?Intestine

Myotome

2 mm

Figure 19 *Cathaymyrus*, our earliest known hemichordate ancestor, from early Cambrian rocks. These little animals, which lived in groups in soft mud, had exactly the same lifestyle as present-day lancelets, simple creatures with the scientific name of *Amphioxus*.
Courtesy of Professor Degan Shu.

Mollusks too were clearly differentiated into their own lineage by the start of the Cambrian. Millions of tiny mollusk shells, mixed with the hard parts of other organisms, are found in the very earliest Cambrian sedimentary rocks. The almost overnight suddenness of their appearance is astounding. Andrew Knoll and his colleagues have found beds of shale in southern China dated just nine million years before the Cambrian that are empty of such fossils, and contain only traces of a few simple algae. Then, nine million years later, these "small shelly fossils" suddenly appeared around the world.

We have few clues about the details of the bodies of the tiny creatures that inhabited these little shells. But the better-preserved fossils from Chengjiang show that 25 million years later mollusks were as advanced in their own way as *Haikouella*. Some of them resembled present-day clams. Others were more like present-day nudibranchs, using toothed mouth parts called radulas to scrub tiny organisms from rocks. The radulas are clearly preserved in many of these fossils.

Some of these early snail-like animals, unlike the nudibranchs, were wildly armored. One of the strangest was the Cambrian creature called *Wiwaxia*. This animal was a total mystery to geologist Charles Walcott, who found the first complete specimens in 1911 in Canadian shale deposits that were laid down twenty million years later than Chengjiang. *Wiwaxia* was oval-shaped, covered with armored plates, and decorated with twin rows of flattened spines that jutted up vertically. It looks like a helmet suitable for a punk rock singer. Walcott thought at first that it must have been a strangely armored marine worm, and later investigators put it into a totally new phylum. But close examination by Simon Conway Morris and others eventually revealed that *Wiwaxia* had a radula-like pair of feeding structures. Argument continues, but it seems likely that the previously mystifying *Wiwaxia* is a kind of primitive and well-armored mollusk.[12]

If these little five-centimeter-long animals crawled along the bottom and scraped their food from rocks, like present-day nudibranchs, then why did they need such elaborate armor? For protection, it seems. There are signs that some *Wiwaxia* shells may have been crushed and damaged by predators before they were buried and fossilized. There were some formidable predators in those Cambrian seas, especially the "fierce crab" *Anomalocaris*.

Figure 20   The mysterious *Wiwaxia*, which turns out to be a bottom-crawling armored mollusk.
© Royal Ontario Museum. Photo: J. B. Caron

Possibly *Wiwaxia* had to resort to armor because it did not have the chemical defenses available to present-day nudibranchs. If, as seems likely, the predators of the time were too stupid to remember which of their prey tasted bad, then chemical defenses would have been useless in any case.

Mollusks may be the exception to the rule that nothing resembling present-day animals left fossil traces before the Cambrian. A small flat animal called *Kimberella*, named after western Australia's ancient Kimberley Range, lived during the Ediacaran period twenty or thirty million years before the start of the Cambrian. There is growing evidence that this puzzling organism had a radula. And it appears to have left scratch marks behind it as it moved across the sea bottom to feed.[13]

The circumstantial evidence is now overwhelming that our ancestors and those of the mollusks had already parted company long before the beginning of the Cambrian. But we do not have the smoking gun—the fossil record of that early divergence has not yet been identified.

Our kinship with the many-talented mollusks

What talented and successful creatures the mollusks are! We are privileged to share the planet with them, and in some modest way to claim kinship with them.

The mollusks have survived by a wide variety of methods. The clams, the mussels, and their numerous relatives protect themselves from predation with strong shells. The cuttlefish and octopuses use jet propulsion to escape from predators. Many of the most highly mobile mollusks lay down decoys of ink clouds and escape in the resulting confusion. Various mollusks have evolved the widest variety of eyes found anywhere in the animal kingdom, eyes that can perceive dangers in their environment in many different ways. Because of their sophisticated eyes, cuttlefish and squid can use elaborate color and luminescence patterns to attract the opposite sex and warn against predators. And, as we have seen, the octopuses and cuttlefish are masters of disguise, using information gathered by their eyes to cleverly match their bodies to their environment.

We chordates, diverse as we are, are boringly predictable compared with the mollusks. Although it is true that vertebrate chameleon lizards can change color, we backboned animals are embarrassingly untalented in other ways. None of us can produce and retract colorful bumps all over our bodies within seconds, like a cuttlefish can (goosebumps don't count).

The most astonishing mollusks of all, the octopuses, have simultaneously evolved high intelligence, the most detailed and complex methods of disguise in the animal kingdom, and the ability to emulate Plastic Man and squeeze through impossible places. It is true that vertebrate snakes can fit through small holes, but they already have a small cross-section. A soft-bodied octopus can squeeze through any hole that is larger than the tiny soft cartilaginous "skull" that lies buried deep within its massive but highly deformable head. A two-foot octopus can squeeze through a one-inch hole!

The differences between ourselves and the mollusks, immense though they seem to be, are beginning to close. Because of convergent evolution our eyes are remarkably similar. And so, in some respects, are our behaviors. Octopuses are the only non-vertebrate animals known to be playful, having often been observed to play with floating objects in aquaria. And, like us, octopuses can quickly open a screw-top jar once they are shown the trick.

When, how, and why did the mollusk lineage part company with ours? We now have some answers to these profound questions.

Comparisons of DNA sequences show clearly that the mollusks and the chordates had a common ancestor. But when these sequences are compared base-by-base, it also becomes clear that many changes—single bases and entire chunks of sequence that have been substituted, inserted, and deleted—have taken place during our divergence from that common ancestor. The accumulation of these numerous differences shows that our common ancestor was remote from us in time. Indeed, that common ancestor lived so far back in time that it is also the ancestor of most of the organisms that I found at Lembeh.

This DNA sequence analysis shows that two major diverging branches of animals arose early from the common ancestor. A sub-branch of one of these two great branches gave rise to the mollusks, along with other important sub-branches that led to the arthropods and to various kinds of worms. The other major branch led to the vertebrates, including us. This second branch also gave rise to further sub-branches that led to—among others— the echinoderms. It is this second major grouping of lineages that sparked my evolutionary musings as I hovered over that spiny mob of fire urchins in the Lembeh Strait. The DNA evidence is unequivocal: the echinoderms are much more closely related to us than the mollusks or the arthropods.

We can put firm dates on some of the events in our ancestry. This is because, whenever we have both DNA evidence and fossil evidence about the ancestry of animals, they tend to agree beautifully. For example, there are a relatively small number of DNA differences between ourselves and chimpanzees—our DNA sequences are 96% identical, and we share almost all our genes. Such a high level of identity tells us that we do not have to travel very far back in time to find our common ancestor. The fossil record agrees with the molecular evidence. Both lines of evidence show that the common ancestor of humans and chimpanzees lived about six or seven million years ago.

The estimates agree so well because our own fossil record is so well studied and because the fossil record of the mammals provides multiple calibration points for the times at which various mammalian DNA divergences began.

But as we move back, to a time before the Cambrian when the fossil record becomes uncertain, it is as if we were walking off a cliff. Without fossils and relying only on molecular evidence, we can be certain that different groups of organisms had a common ancestor, but we are not sure when that ancestor lived. The further back we go, the greater the uncertainty over timing grows.

Part of this is the fault of the DNA sequences, which can diverge at different rates. For example, when Kevin Peterson of Dartmouth University and his colleagues built evolutionary family trees using vertebrate and mollusk DNA, they found that the vertebrate branches of the tree were only half as long as the mollusk branches. It seems that vertebrate DNA in general has evolved at only about half the rate of mollusk DNA. We do not know why these rates are different, and why these different rates have been maintained for well over half a billion years.

Despite these difficulties, many groups of scientists have used the growing library of DNA sequences to probe the distant past. They have been cautious, extrapolating back from well-supported dates, and using a variety of assumptions about the rates of DNA evolutionary change. Some of these studies estimate that the common ancestor of scuba divers and cuttlefish might have lived as much as a billion years in the past. Others, using different statistical methods, arrive at the more recent date of 650 million years ago, a mere hundred million years before the start of the Cambrian. Still other estimates fall between these extremes. But none of the dates are so recent that they fall within the Cambrian itself. The consensus is that these animal lineages did indeed begin to diverge at some point in time well before the start of the Cambrian.[14]

Thus, by the beginning of the Cambrian, the fossil and DNA evidence agree that much diversification had already taken place, though the exact nature of that diversification remains a mystery. At the time of the Cambrian explosion, environmental changes allowed each of these already-divergent lineages of small soft-bodied organisms to grow larger and evolve various hard parts such as skeletons or shells, so that they were more likely to be fossilized. Thus, the Cambrian explosion was not really sudden. In the memorable phrase of Simon Conway Morris of Cambridge University, it had a long fuse.

How we and the mollusks first parted company

The DNA evidence makes clear that my feeling of kinship with the mollusks of Lembeh is well founded. But how and why did we first part company, and why did we take such separate evolutionary paths? These questions are much harder to answer, because they require evidence from the fossil record that we do not yet have.

Some intriguing hints of such evidence come from Precambrian shales and carbonate rocks in the Doushantuo formation of Guizhou Province in southern China. These beds, which date from 40 million years before the Cambrian, have yielded some tiny but well-preserved fossils that look like dividing cells. They might, as their discoverers suggest, be the remains of simple one-celled animals with cells like ours.

Tiny vase-shaped fossils from the same beds look as if they might have been the embryonic stages of small animals. But these embryos, if that is what they are, are not accompanied by any signs of the animals that they might have grown into. Nowhere in the Precambrian rocks have scientists yet discovered anything that looks like the later Cambrian organisms, aside from fragmented glassy skeletons of early sponges and that enigmatic proto-mollusk *Kimberella*.

Why were the ancestors of the Cambrian organisms so small and soft-bodied that they left such sparse and enigmatic fossils? Andrew Knoll suggests that two things may have happened. First, oxygen levels gradually rose to the point at which larger organisms could have been supported. But at the outset this rise may have had little effect, because the ecological niches of the Precambrian world were already full and there were simply no opportunities for such large-bodied creatures. At the beginning of the Cambrian, however, there is evidence for an extinction event. The resulting wave of extinctions emptied ecological niches everywhere on the planet. These new opportunities for life, coupled with the more plentiful oxygen, may have been the trigger for the Cambrian explosion.

If this hypothesis is correct, then the tiny ancestors of mollusks, worms, arthropods, and chordates that lived at the beginning of the Cambrian and survived the extinction event were finally able to take advantage of high oxygen levels and evolve into big active animals in an empty world where new ecological opportunities abounded. The oxygen-extinction hypothesis has the virtue that it explains the suddenness of the Cambrian explosion, while a scenario that relies solely on a gradual increase in oxygen levels does not. In Chapter 5, we will see evidence for a similar explosive takeover of ecological niches as mammals took over from the dinosaurs, aided in their efforts by newly emerging properties of flowering plants.

How to start on new evolutionary paths

The remaining questions that confronted me in Lembeh were perhaps the most profound. What actual physical changes took place in the bodies of our ancestors when they parted company with the ancestors of the mollusks? And might it be possible to recreate, in present-day laboratories, some approximation of those ancient changes?

Let us begin by looking at the genes that control how animals look and behave, because it is such genes that must have been involved in those dramatic Precambrian events. These genes, known as regulatory genes, govern our development from embryo to adult. They control the time and place at which other genes are switched on and off. And it is the regulatory proteins coded by these genes that must hold the answer to the shape-shifting that took place during the early divergence of multicellular life.

When regulatory genes are damaged by mutations, the results can be profound. Such developmental changes can affect the entire organism as it grows and matures. The consequences are sometimes dramatic and grotesque. In fruit flies, some of these mutations produce flies with four wings rather than the usual two, or legs that grow out of their heads instead of antennae.

My colleague Marty Yanofsky has produced equally extreme regulatory mutants in plants. He has made mutant forms of the little wild mustard plant *Arabidopsis*. These mutants have flowers in which all the different parts have been converted to sepals, the outer leaves of the flower bud. If such mutant plants had appeared in nature rather than in Marty's laboratory they would have died without reproducing.

Susan Lindquist and her colleagues at the Whitehead Institute in Cambridge, Massachusetts, have gained further insights into mutations that have such large effects by working with the fruit fly, *Drosophila melanogaster*. Under the microscope this tiny fly is revealed to be a complex and jewel-like creature, with bright red faceted eyes, subtly patterned wings, and antennae that allow it to home in on the tiniest chemical signals from that slice of cantaloupe you are eating.

Lindquist and Suzanne Rutherford examined fruit flies that make a defective form of a protein called a chaperonin.[15] Chaperonins are proteins that, as their name suggests, act as guardians of other proteins.

Many of the proteins in our cells are extremely fragile. When they are being synthesized by the cellular machinery they tend to flop around like newborn babies. Chaperonins bind firmly to these delicate proteins during the critical birth process, coercing them to take the right shape so that they can play the correct role in the cell's development. Like Mary Poppins, the chaperonins permit no nonsense from their unruly young charges. They make sure that their protein pupils, many of which play important regulatory roles, get to their proper place in the cell and bind to the right parts of other proteins or to the right regions of DNA, without the molecular equivalent of making funny faces in the process.

The chaperonin that Rutherford and Lindquist investigated is a "heat-shock protein" called Hsp90. High temperatures are dangerous to proteins, and many organisms, ourselves included, synthesize plentiful amounts of these heat-shock chaperonins to protect our other proteins under these extreme conditions.

Fruit flies cannot survive if they make no Hsp90, so Rutherford and Lindquist used flies that carried one damaged and one normal form of the Hsp90 gene. These mutant flies made half as much of this chaperonin as normal flies. This small genetic change was enough to cause a few of the flies to develop abnormally. Among the many different kinds of abnormalities, some of these flies had misplaced and misshapen eyes, while others had wrinkled wings.

When Rutherford and Lindquist picked some of these abnormal flies and bred them, they found that within a few generations all the progeny were abnormal. Even the progeny flies that had two functioning Hsp90 genes continued to show abnormalities.

What had happened? When Rutherford and Lindquist saw funny-looking flies among the progeny of their crosses, they were finding the flies that had the least robust developmental pathways and that were therefore most likely to be sensitive to low chaperonin levels. After they had selected and bred these flies for several generations, they ended up with lines of flies with developmental pathways that always tended to be easily disturbed. The flies' development was abnormal even when chaperonin molecules were present in their usual numbers and were doing their best to maintain discipline.

Rutherford and Lindquist's developmentally disturbed fruit flies had so many things wrong with them that they could never have survived outside of the laboratory. Such organisms would probably not have survived the kinds of large developmental disturbances that might have sent Precambrian creatures off on new evolutionary paths. But perhaps the criteria for survival are less strict for animals smaller and simpler than fruit flies. Small life forms that consist of only a few cells, such as the early animals that lived in the Precambrian, might have a better chance of surviving drastic body-plan modification than large complicated organisms such as present-day fruit flies.

What if we could perform experiments like those of Rutherford and Lindquist on organisms simpler than fruit flies? How drastically could we modify such simple organisms and still leave them able to survive and even thrive?

There are signs in present-day animals that drastic modifications of their remote ancestors' body plans did indeed take place. Early in the nineteenth century the French anatomist Geoffroy Saint-Hilaire came to a remarkable conclusion about a major pair of branches in the tree of life. Vertebrates, he declared, are simply upside-down arthropods (or vice versa). At some point early in our history we (or they) flipped over, so that our spinal cords form along our backs and those of the arthropods form along their bellies.

However, our faces and the faces of arthropods do not show this rotation. Both groups of animals have their eyes above and their mouths below. So it may be that the first Precambrian ancestor to undergo the flip did so by a feat of contortion worthy of the Boneless Wonder in a circus sideshow. Its body rotated 180° behind its head, leaving its head in the original orientation.

Saint-Hilaire's explanation of the body plan difference between arthropods and chordates, widely ridiculed at the time, has turned out to be correct.[16] It is of course difficult to imagine such a drastic rearrangement happening in stages. And the rearranged organism was more likely to have survived if it was a small simple Precambrian creature than if it was a larger and more complicated creature living at some later point in time.

Recreating the Precambrian

We cannot yet recreate the drastic developmental mutations of the Precambrian in the laboratory, because we have no Precambrian organisms to experiment on. But it is not beyond the realm of possibility to make and study similar changes in simple laboratory organisms available today.

A good candidate for such experiments is the tiny roundworm *Caenorhabditis*, which is a mere one millimeter long. This worm normally lives in soil, but it can easily be raised in the laboratory. And its development is delightfully simple and predictable. An adult worm's body is made up of exactly 959 cells, no more and no less.

Caenorhabditis has genes for Hsp90 and other chaperonins. It is now possible using molecular techniques to "knock out" this and any other of the genes of these little worms. It has been found that damage to the Hsp90 gene causes problems with the worms' metabolism and shortens their lives.

Suppose that we damaged these and other chaperonin genes in the worms and placed the resulting mutants in a variety of new environments? Would it be possible to select for worms with a different body plan? Could some of these changed worms survive, and even thrive, under the altered conditions that we impose on them? Perhaps we could produce changes in these worms that are as far-reaching as the dramatic reorganizations that happened to our tiny ancestors during the Precambrian, more than half a billion years ago.

Something similar has been done using computerized life forms. In 2000, Hod Lipson and Jordan Pollack of Brandeis University carried out an important study of such cyber-creatures.[17]

Lipson and Pollack created a diverse population of replicating life forms in a computer, and competed them to see which could move most quickly across a virtual flat surface. These virtual life forms were simple. They were controlled by a collection of virtual electronic circuits representing a minimalist "brain." The brain circuits were connected to a variety of virtual body parts such as rods and ball joints. The brains of these computer creatures could move the rods and cause them to change their length, and could rotate the ball joints that linked these rods together.

These "organisms" were allowed to replicate themselves in the computer. The computer program that directed their replication was instructed to introduce occasional random changes, so that mutant organisms arose each generation. At random, the brain circuits could be switched to new patterns, the linkages between the various rods and joints could change, and body parts could change their character from one type to another. Because the mutations happened at random, just as in the real world, most of the mutant organisms were grotesque constructs that received random signals from their "brains" and flailed about uselessly. A minority of them, however, could do something useful.

The computer program then dumped each generation of organisms onto a virtual flat surface, and monitored how quickly they could crawl, hump, writhe, or wriggle their way across it. The slowest creatures were condemned to cyber-oblivion, and the fastest creatures were allowed to replicate and to undergo further random mutations.

In experiment after experiment, selection for the fastest organisms resulted in the emergence of certain types of body plans. One especially effective type was a little rigid pyramidal shape enclosing an angled rod that could change its length. The little creature's brain was wired to drive the rod repeatedly down and backwards, sending it across the level surface like a pole-driven punt on the River Cam. Another extremely efficient creature was shaped like an arrow. It "rowed" itself forward by two extensible arms set at angles to its "head," just like a little rowboat. A third creature, a twisted parallelopiped,

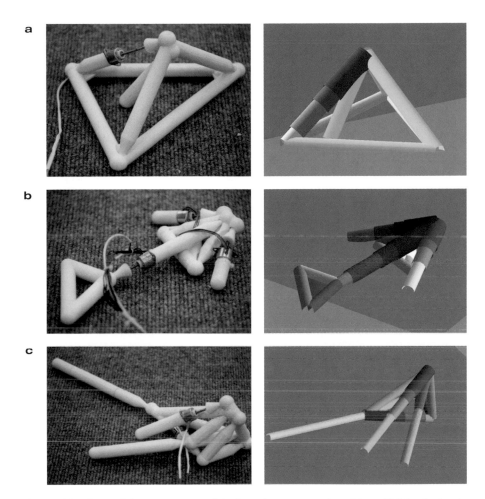

Figure 21 Some of the more successful robots that were produced by artificial selection in the computer world of Lipson and Pollack. The creatures generated by the computer program are on the right, and on the left are real models of them that turned out to be equally successful in moving across a flat surface. Note that creature (b) is clearly bilaterally symmetrical, a body plan that took minutes to emerge in the computer but may have taken millions of years to emerge in the Precambrian seas! (From Figure 5 of Lipson and Pollack, *Nature* 406 (2000): 974-8.) © Hod Lipson and Jordan B. Pollack, Brandeis University

hobbled along by extending and contracting itself like a distorted inchworm. When the experimenters constructed real models of these virtual organisms, the models were able to move swiftly across a real surface.

One striking feature of this strange menagerie was how often these virtual creatures "evolved" a head and a tail end and bilateral symmetry. Like our own bilateral body plan, their left and right sides were sometimes mirror images of each other. Convergence on this body plan happened repeatedly, even though the cyber-creatures' ancestors had started out as different random and ineffectual collections of parts.

It is perhaps not coincidental that one of the earliest events in the evolution of animals, an event that happened long before the Cambrian, was the emergence of bilateral symmetry from ancestors that had previously been round or irregular in shape and that lacked a head or a tail. Bilateral symmetry gives organisms the ability to move directionally through the environment. It seems to be a highly favored evolutionary path, both in Precambrian organisms crawling across a mud sea bottom and in virtual organisms that must make their way across a computer-generated surface.

Lipson and Pollack's experiment provides us with some guidance on how to design our *Caenorhabditis* experiments. Because this worm is already bilaterally symmetrical, we will have to select for other types of body-plan changes. Perhaps collections of mutated *Caenorhabditis* could be selected for the ability to wriggle quickly across a smooth surface, while at the same time they are being buffeted by a current of water flowing in the opposite direction. Would we select for worms that can adhere to the surface, so that they can wriggle forward despite the current? Or would we select for worms with sail-like structures that would allow them to tack against the current? The possibilities are endless. Perhaps we could create a world like the Precambrian one, in which it is possible to select for many different body plans simultaneously.

Closing the gap between scuba divers and cuttlefish

As I roamed the floor of the Lembeh Strait I met many radially and bilaterally symmetrical creatures that are the remote descendants of Precambrian developmental mutants. When I locked gazes with my inconceivably distant relative, the flamboyant cuttlefish, I felt a kinship that reached across the 600 million years of accumulated genetic differences that separate us.

Cuttlefish and octopuses are the world's most expert shape-shifters. We have lost that ability, but our remote ancestors and theirs were shape-shifters too. They underwent changes in shape as they evolved in the Precambrian seas, and it was those alterations that set us on our different paths. And now we are on the verge of recreating and understanding such changes.

As I hovered in the magic world of Lembeh another question occurred to me.

If we and other vertebrates were to go extinct, leaving the field open for octopuses, could these organisms too develop culture and science? Could an intelligent and daring octopus eventually propose a theory of natural selection? And perhaps other octopuses, offended by the presumptuous scientist's attack on the octopus god that created them, would exclaim: "Nonsense! We could not possibly be descended from that ugly *Wiwaxia* creature!"

2

The Inner Workings of Evolution

Figure 22 At flood, the rivers of the upper Amazon basin overflow into lakes that line their shores. Here a lake near the Marañòn River, covered with edible water lettuce *Pistia stratiotes*, provides an ideal habitat for Hoatzin birds. In this chapter we will explore how the evolution of these astonishing birds has cast strong light on how evolution really happens.

In the previous chapter we visited some of the further branches of the great evolutionary tree of life. We asked whether it might be possible to recreate, or at least imitate, some of the early evolutionary changes that took place down near the base of these branches, changes that eventually led to the diversity of present-day animals.

It is clear that the Darwinian process of natural selection must have been the primary force that drove those early events, but natural selection is only a part—albeit a most important part—of how evolution happens. In this chapter we will examine some further examples of evolution in action, in places ranging from the heat-blasted soils of Israel's Negev Desert to the dense mangrove forests of southern Borneo. In the process we will explore in detail how evolution actually takes place.

Evolution Canyon—a Darwinian laboratory

Recently a group of us, fresh from a meeting at Haifa University to commemorate the bicentenary of Darwin's birth, accompanied the evolutionary biologist Eviatar Nevo on a field trip. He took us to his favorite research site in northern Israel and gave us a demonstration of how evolution is taking place before our very eyes.

Nevo, a lively snowy-haired octagenarian with an overwhelming passion for science, is the founder of Haifa University's Institute of Evolution. He strode confidently into the canyon that he has studied for forty years. The canyon, officially called Nahal Oren, is one of several that run east and west along the slopes of Mount Carmel, less than an hour's drive from the spectacular coastal city of Haifa. Here the air is fresh, birds sing, and oleanders and Judas trees bloom on the valley's sunny floor.

The northern and southern slopes of the valley differ dramatically. On the wetter and more shaded north-facing slope the vegetation is plentiful, with scrub oaks, wild olives, and other small trees predominating. Just a few hundred meters away across the valley floor is the sun-drenched south-facing slope, a rocky terrain covered with dry grasses and dotted with acacia shrubs. The north-facing slope resembles southern Europe's Mediterranean

Figure 23 Eviatar (Eibi) Nevo stands on the shaded "Mediterranean" side of Evolution Canyon in northern Israel. He is explaining the large evolutionary differences that he and his colleagues have found in the animals and plants living on this shaded slope and members of the same species on the sunny "African" slope behind him. You can also see behind him one of the many caves that dot the slopes of Mount Carmel. Some of these caves were inhabited by Neanderthals and early modern humans over spans of tens of thousands of years.

coast, while the south-facing slope transports us to the dry African savanna. Here, in what he calls Evolution Canyon, Nevo has built his career on the study of an ecological collision between two continents.

For the past forty years, Nevo and hundreds of his students and colleagues have examined the animals, plants, fungi, and bacteria that populate the canyon. They have revealed an evolutionary tapestry in which environmental differences, natural selection, gene mutations, and the ebb and flow of genes within populations have shaped every aspect of the canyon's life.

Some plants and animals are confined to only one of the two slopes of the canyon, but many others such as fruit flies and butterflies roam the entire canyon. In virtually every case the physiology of the animals and plants reflects their location. Those on the hotter and sunnier African side are more resistant to water stress, and in this bright environment many animals and even some fungi secrete more of the dark UV-protective pigment melanin.[1]

These differences in physiology between the populations of organisms on the two slopes of the canyon are reflected in their genetic composition as well. One of the great strengths of modern evolutionary biology is that it has given us the tools to plumb the depths of the gene pools of every plant and animal species, including our own. We can detect and measure genetic differences, even those between adjacent populations that look superficially alike.

The gene pool of a species is far vaster and more complex than the genotype of an individual of that species. The genes in this collective pool come in many different forms, called alleles. For each of our genes we as individuals can carry at the most two alleles, one from each of our parents. The gene pool of our species, in contrast, may contain dozens of different allelic forms of the same gene. To get an idea of the range of variation that our gene pool contains, we must look at many different individuals. Gene sequencing and other methods now allow us to measure the variation within our own population and within the populations of other animals and plants.

Dissecting the process of evolution

Eibi Nevo has applied this approach to the species of Evolution Canyon. He and his students and collaborators have peered within the gene pools of many different populations that live in the canyon. In species after species they have found that the populations living on the two sides of the canyon show differences in allele frequency at many different genes. These differences parallel the different physiological, behavioral, and color differences that these populations exhibit (though in most cases the genes that actually determine these physical differences are not known).

Some of the genetic differences have resulted from natural selection for and against different alleles on the two slopes of Evolution Canyon. Natural selection shifts the frequencies of alleles in gene pools, and it is these shifts that bring about evolutionary changes in populations.

The natural selection acting in the canyon must be strong, because it is able to maintain the genetic differences between these adjacent populations. These differences are maintained even though many of the animals can fly, crawl, or scuttle freely from one side of the valley to the other, and of course the pollen and seeds released by the plants can blow everywhere.

Because of these movements, genes from both sides of the canyon mix every generation. Even a limited amount of such gene flow should be enough to wipe out any genetic differences on the two sides of the canyon. The members of a species should look and behave the same on each side. Nonetheless, the populations are different. Selection preserves and reinforces the genetic differences between the populations even though there is copious gene flow between them.

Sometimes the effects of this selection can be measured directly. Two common species of Drosophila fruit flies live in the canyon. In both of these species, Nevo's colleague Abraham Korol has found a strong tendency for the flies to choose partners from the same slope.[2] The flies have been selected to detect differences between potential mates from the two sides of the valley. This tendency to mate non-randomly preserves the genetic differences between the gene pools on the north and south slopes, even though the flies in theory are all perfectly capable of flying anywhere in the canyon and mating with whichever fly they choose.

Korol's work illustrates an important aspect of natural selection that is often overlooked when we talk about evolution. Natural selection is not just about selecting new mutations. Natural selection can, and often does, act to *preserve* the genetic differences between populations. It can also preserve the genetic variability hidden within populations, which makes the variability available for later evolutionary changes.

Korol and Nevo revealed another essential part of the network of evolutionary forces operating in the canyon. Fruit flies that live on the hotter, brighter side of the canyon are able to repair damage to their DNA more readily than those that live on the cooler side.[3]

Mutational changes in the DNA are the ultimate source of genetic variation. The many allelic forms of genes in gene pools were all originally produced by mutations. Without mutations evolution would grind to a halt, for there would be no genetic variation on which natural selection could act.

Mutations can be caused by many things, including radiation and the invasion of our cells by viruses that can change our DNA. Most mutations, however, arise from mistakes that occur when DNA is copied. Whenever a cell divides into two daughter cells, all its DNA must be replicated. The enzymes that do the copying are really good at their job, but they are not quite perfect. This is not surprising—making a perfect copy of all your DNA would be the equivalent of copying out *War and Peace* a thousand times without a mistake.

There is now strong evidence that organisms living in a stressful environment have a higher rate of mutational mistakes than those living under more relaxed conditions. To compensate, they also have better mechanisms for repairing their DNA if it is damaged or if mistakes are made during the copying process. Korol has found that *Drosophila* on the sunny side of the canyon, which are exposed to thermal stress and blasts of ultraviolet radiation, are able to repair their DNA more readily than flies that live on the shadier side. These genetic differences between the two fly populations have also been maintained in the face of gene flow between the two sides of the canyon.

Our current understanding of evolution has been much refined and extended since Darwin's time. We now know that the composition of gene pools changes because of natural selection, mutation, chance events, the flow of genes among populations, and the powerful mixing process of genetic recombination that accompanies sexual reproduction each generation. Towering over all, and driving these evolutionary changes, are the interactions between organisms and the environment in which they live.

As Nevo roamed around the canyon, he pointed out mounds of fresh earth that mark the burrows of some remarkable animals that have also been shaped by these processes of evolution. In the canyon the burrows house the mole rat (*Spalax ehrenbergi*), a blind naked mammal that lives all its life underground. There are more than a dozen species of *Spalax*, four of which are found in Israel.

Some species of mole rat that live in Africa have lifestyles that resemble those of social insects such as termites. One female dominates each network of tunnels and produces all the babies, while the other females and the males bring her and the babies food and defend the tunnels fiercely. The Israeli mole rats, separated by deserts and the Red Sea from their African relatives, live more solitary lives, and all the females are able to reproduce.

Even though the mole rats live underground, their evolution has been strongly influenced by climatic conditions in the world above. Each species occupies a different Israeli ecosystem. A species with 60 chromosomes lives in the hot dry south, including the sandy wasteland of the Negev Desert. A species with 58 chromosomes is found in the more climatically moderate region around Haifa and the Sea of Galilee. And two other species with smaller numbers of chromosomes thrive in the hills of the Golan Heights.[4] These northern areas are being fought over by their human inhabitants in part because they receive fifteen times as much rain as the Negev. The mole rats, meanwhile, burrow peacefully under Israel's dangerous northern border with Lebanon and Syria, blithely ignoring the politically charged world above them.

Nevo has shown that rainfall and the permeability of the soil to oxygen are tightly correlated with the distributions of these species. The boundaries between the species are sharp. Matings between the mole rats at the boundaries produce some hybrid animals, but these hybrids are ill-adapted and do poorly.

Chromosome differences probably have something to do with the barrier to the free flow of genes between the species. Intriguingly, the zones in which hybrids are found are relatively wide when there are few chromosomal differences between the species. They are narrower and more constricted, indicating stronger selection against gene flow between the species, when the chromosomal differences are greater. The different species of *Spalax* have evolved through the same processes of mutation, selection, genetic recombination, and gene flow that are operating in Evolution Canyon.

Perhaps, given enough time, the environmental differences between the north- and south-facing slopes of Evolution Canyon might be able to generate new species. Korol argues that this will not happen, because the differences between the two sides of the canyon are not huge and gene flow is able to mix the gene pools at a high rate. But I contend that if gene flow is reduced further in the future, some species in the canyon might split into two.

Species like the burrowing *Spalax*, which do not migrate long distances, might be easily split if barriers are put in their way. Already, a new road that runs through part of the canyon is introducing stronger barriers to the free exchange of genes. Mole rats cannot burrow under the road, because they depend on oxygen that diffuses from the surface. The road's impermeable tarmac prevents this diffusion. Other, above-ground species may not be able to move quickly enough to dodge Israeli drivers. Of such small changes is most evolution made.

Nevo's work in Evolution Canyon, along with similarly detailed evolutionary scenarios that have been uncovered by thousands of other scientists in every part of the world, have given us profound insights into how evolution happens. This new work goes far beyond the early insights of Darwin, who knew nothing of genes or gene pools. We now know that evolution is a dynamic process in which changes in the composition of gene pools are driven by natural selection. Severe environments increase the effectiveness of natural selection. They may also introduce more mutations, the ultimate raw material of evolution.

Nevo has spent his life studying how changes in the genetic structure of populations can be linked to ecological differences. At the end of our visit to the canyon he encouraged our little group of scientists and students to collaborate in his ongoing studies. His twinkling enthusiasm has inspired generations of students from around the world to come to Evolution Canyon so that they can find out how evolution really happens.

Eats shoots and leaves

The study of how genes have evolved has revealed a profound truth about how evolution works. Evolution rarely starts from scratch. Instead it utilizes genes and genetic pathways that have already been shaped by natural selection to solve a different adaptive problem.

Evolution boils down to a matter of probabilities. It is quite likely, for example, that an extra copy of a gene will be inserted into a genome by mutation. Once the new copy of the gene has appeared, it can evolve to take up a new function. This is a far more likely sequence of events than the possibility that a gene for the new function will arise *de novo* through mutation.

As an example of how evolution tends to take the more likely path of building on pre-existing functions, let us see how hoatzin birds of South America and the proboscis monkeys of Borneo have solved a particular evolutionary challenge in similar ways.

The challenge is a formidable one. Somehow, these animals can survive by eating the leaves of trees, even though they are unable to digest them! Let us begin by meeting the fascinating cast of characters in this evolutionary drama.

The first character is the South American hoatzin bird. The hoatzins are unique among birds, not just because they have carved out this extremely specialized dietary niche, but because they are living fossils, the last remnant of a unique bird evolutionary lineage. They are so astonishing in so many ways that it will be most instructive to spend a little time with the hoatzins.

Raucous hoatzin colonies are common along the rivers of the Amazon basin, and they are best seen from a boat. I first encountered these cheerful if feckless birds on the margins of the Manù River, a tributary of the upper Amazon, in a relatively undisturbed part of the vast rainforest that lies east of the Andes in Peru.

Hoatzins are the punk rockers of birds. They squawk enthusiastically and have a wild hairdo of untidy feathers. They also live in a dangerous world in which their babies must often fend for themselves against predator attacks. Luckily the babies are up to the task, because they are able to draw on ancestral capabilities that go back to the beginning of the Age of Dinosaurs.

When threatened by predators, haotzin babies jump out of the nest to the ground or even into the water. For most bird hatchlings this would be a sentence of death. But these babies can crawl back up to their nest using a pair of claws on each of their wings. As they mature into adults and learn to fly, they lose their claws.

The babies' claws, unique among present-day birds, resemble the wing claws that were possessed by *Archaeopteryx*, an early bird that lived 150 million years ago. The claws of *Archaeopteryx* were in turn remnants of the claws of the birds' dinosaur-like ancestors. The *Archaeopteryx* claws, three on each wing, were lost in the course of subsequent bird evolution.

Figure 24 A pair of hoatzins, *Opisthocomus hoazin*, peer from their perch above the *Pistia*-covered upper Amazon lake. Hoatzins eat leaves, including water lettuce leaves, almost exclusively.

But were they? Claws have reappeared in the hoatzin lineage. Have the hoatzins resurrected an ancient developmental pathway that has lain hidden and inactivated in bird genomes for 150 million years? Are they survivors of a bird lineage that retained the claws during all this time? Or have they evolved a brand new developmental pathway? We are not yet able to distinguish among these possibilities, though the last possibility is the least likely one. There is growing evidence that it is far easier to restore the genetic functions of an old suppressed developmental pathway than it is to evolve a new pathway from scratch.

We do not know whether some of the hoatzin's near relatives might also have had wing claws, because all of these near relatives seem to have been lost through extinctions. Argument continues about whether hoatzins are more closely related to the doves, the cuckoos, or the turacos, such as the

Grey Lourie or go-away bird that I surprised in Botswana (see Figure 25). The DNA evidence is equivocal. The relationships between the hoatzins and other present-day branches of the avian family tree are extremely distant, making the exact affinities of the hoatzins difficult to decide.[5]

Even though all their close relatives have disappeared, hoatzins are quite common birds. Luckily for them they taste rather awful and are not hunted by the local tribes. One of the reasons for their foul taste is that they dine almost exclusively on the shoots and leaves of about fifty different plants. Their bodies are loaded with breakdown products of the tannins and cardiac glycosides that the plants secrete to repel insect and animal predators.

The hoatzins must eat a wide variety of plants in order to keep the various toxins at manageable levels. This mush of different plant tissues, each quite indigestible by the hoatzins themselves, is transformed into nourishing compounds inside their foreguts by a process of fermentation.

Figure 25 This grey turaco or grey lourie (*Corythaixoides concolor*) is known as the go-away bird because its song sounds like it is demanding that you leave immediately. I photographed it in Botswana's Okavango Delta. It may be one of the closest living relatives of the hoatzin, but close is a relative term, for hoatzins and turacos occupy distinct branches of the avian evolutionary tree.

Figure 26 A hoatzin launches clumsily into the air. Swollen crops make these birds poor flyers—they flap through the air like galleons under full sail.

Animals that ferment their food, such as hoatzins, sheep, and cows, eat grass or leaves in enormous amounts. They must then subcontract the job of digesting this mass of plant tissue to a carousing band of bacteria and other single-celled organisms that inhabit their stomachs.

Eat, drink, and be merry

The second member of our cast of characters is a collective. It is the thriving colonies of bacteria that carry out this fermentation and that live in symbiosis with their hosts. Some of the bacterial species are so specialized for this mode of life that they are found nowhere else. Hosts and bacteria absolutely depend on each other.

Some birds, such as grouse and ostriches, carry out a little fermentation in various parts of their guts, but no other living birds are able to ferment the food in their crops as effectively as the hoatzins.[6] Hoatzins can retain the fermenting mix in their crops for long periods, an unusual ability for such small-bodied animals, and as a result they are just as efficient as sheep in gaining energy from the fermentation process. When did they acquire this fermentative expertise? We do not know, just as we do not know when and how their babies acquired their claws, because all the close relatives of the hoatzins have disappeared.

As a result of their symbiosis with digestive microbes, the crops of hoatzins are greatly swollen. This gives them a comic embonpoint and makes them awkward fliers. And this in turn is why they build nests on riverbanks rather than in the depths of forests—they need a clear line of flight to land on their nests.

The energy that the hoatzins depend on comes from an apparently unpromising food source: plant fibers. Woody plant tissue is chiefly made of cellulose. The fibers of cellulose in turn consist of long chains of the sugar glucose. In theory, therefore, cellulose should be eminently digestible. Unfortunately, its glucose molecules are bound together by chemical linkages that need special hydrolytic enzymes to unlock them. The plants' woody fibers have been made even less palatable through a wide variety of chemical modifications, including tight chemical linkages that bind the cellulose to other carbohydrate chains called lignins and hemicelluloses. The result is a thicket of interlocking molecules. These strong woody fibers are superb at supporting the plants so that they can thrust their leaves into the air and sunlight. At the same time the fibers are totally resistant to the digestive systems of insects, birds, and mammals.

Enter these new members of our cast of characters, the fermentative bacteria, which are able to break down these fibers. The animal hosts of these bacteria have evolved many ways to help them in their task.

The most sophisticated ruminant digestive systems, such as those of cattle and sheep, have stomachs that are divided up into a series of four chambers. Recently swallowed food is easily regurgitated back into their mouths from the first two chambers. The animals can masticate their food and mix it with saliva whenever they have the leisure to safely chew their cud. The softened and broken fibers are then swallowed again, so that they are more easily fermented by the huge populations of microbes that live in these first two fermentative chambers.

The overflow of this burgeoning population of microorganisms then spills over to the last two stomach chambers. There the nutritious organic acids and fatty acids that the microbes produced during the fermentation stage are absorbed by their host.

But wait, there's more! The microbes had a high old time as they dined on all that free food, but now they must pay for their brief Lucullan orgies. When they reach the acid-filled stomach chambers they too are digested, and their component parts are absorbed by the ruminant's stomach and small intestine.

The lives of these bacteria are short but (one hopes) happy. And these characters in our drama have also been shaped by evolution. Ruminants that have become specialized to eat leaves or grasses carry different specialized bacterial populations with which they have co-evolved. Each host carries the right mix of bacterial species to extract nourishment quickly and efficiently from their particular type of food.

Hoatzins do not have quite as sophisticated a stomach as the ruminant grazing animals like cows and sheep. But they do have a two-chambered crop, the first chamber being an expanded region of the lower part of the esophagus that leads into the stomach. Most of the fermentation takes place in the hoatzin's swollen lower esophagus, and most of the digestion in its stomach.

A primate fermenter

The third member of our cast of characters is a close relative of ours, the leaf-eating—and leaf-fermenting—proboscis monkey. My most dramatic encounter with these primate fermenters took place in a vast area of mangroves and wet rainforest in Kalimantan, the southern part of Borneo.

This Indonesian province comprises the southern three-quarters of the island. It is the wildest and at the same time the most endangered part of Borneo. Its coasts are home to vast regions of wet lowland forest, and its great rivers nourish thriving mangrove ecosystems. Both the wetland forests and the mangroves are threatened by uncontrolled development and logging.

We hired a flat-bottomed boat to take us up the Sekonyer River, to visit the Canadian primatologist Birute Galdikas at her orangutan research station. On the way we surprised a troop of proboscis monkeys in a tree overhanging the river. The troop we encountered was a collection of several male-dominated harems. Surplus males who do not have harems band together into separate (and presumably frustrated) single-sex troops.

These monkeys, even the babies, have swollen stomachs, signs that they are foregut fermenters like the hoatzins. The faces of the female and immature proboscis monkeys resemble those of their close relatives the langurs, but the males have enormous fleshy red noses that droop down over their mouths and that swell when they are angry. When this happens they look like bad-tempered, overweight comedians. As they scream at observers they sometimes pile on another level of threat by displaying bright red erections.

Proboscis monkeys, like many primates, are endangered by habitat loss. They are now found only in the lowland river systems of Borneo, though they were almost certainly more widely distributed in the past. Like the hoatzins these animals are ecological specialists. At the extreme, evolution can sometimes lead to adaptations so precise that they cut off any avenues of escape if the environment changes. This seems to have happened to the proboscis monkeys. They cannot even be raised in zoos, because they soon waste away and die in captivity.

Two thirds of the diet of proboscis monkeys consists of leaves. They spend most of their time in coastal mangrove forests, where leaves of mangrove and pedada trees are plentiful. Some of them do spend part of their time in the rich and diverse forests that lie just inland of the mangroves, where many different kinds of fruit are plentiful. But they tend to eat only unripe fruits, ignoring the ripe and sugary fruit that abound in the forest. The bitter, tannin-rich mangrove leaves are their chief source of food, except during seasons when fruit is plentiful.[7]

Why do these monkeys confine themselves to such a monotonous and—at least to an omnivorous primate like myself—apparently unappealing diet,

Figure 27 (*opposite*) This male proboscis monkey, *Nasalis larvatus*, is on the Kinabatangan River in Sabah, Borneo. Notice his fermentation-swollen stomach.

Figure 28   A moment later he sucked in his stomach and erupted in rage at our boat.

when the plentiful riches of the rainforest lie all around them? A likely explanation is that proboscis monkeys split from other colobus monkey lineages
primarily because they acquired the ability to eat mangrove leaves. This gave
them exclusive access to the immense resource of the mangrove forests,
which covered a far larger area than they do today. Once this monkey lineage
was able to move into the mangrove forest's maze of swamps and streams it
was also largely protected from predators such as the clouded leopard.

These monkeys have a spectrum of talents that allow them to flee any
predators that do manage to penetrate the mangrove swamps. As we watched,
the troop showed off its natatorial skill. One by one the monkeys dived gracefully from high branches, springing as far out as they could to land in the
middle of the river. Once in the water they surfaced and swam rapidly to the

Figure 29 In an image captured by my daughter, a proboscis monkey dives into Kalimantan's Sekonyer River. You can see another monkey waiting its turn on the branch. (Photograph by Anne-Marie Wills Rivera.)

other bank. Their swimming style was impressive, a swift dog paddle with an occasional overhand stroke that reminded me of the Australian crawl.

Swimming speed is essential for these monkeys, because the rivers of Kalimantan are full of crocodiles. And their ability to swim is more than some frantic splashy attempt to escape danger. It has been refined through a long period of natural selection. The webs between their fingers and toes are more extensive than those of any other primate.

Like the hoatzins, the proboscis monkeys have highly specialized digestive systems. Fermentation begins in their foregut chambers from the very moment the food is swallowed. As with other fermenters, a small change in their diet can lead to a wide range of problems. They must limit their intake of sugar-rich fruits, because an influx of sugar would be enough to disturb the bacterial ecosystem on which the fermentation depends. Similar disturbances in bacterial ecology can lead to the often fatal conditions of acidosis and bloat in domestic cattle.

These monkeys' specialized diet may be why they cannot be raised in captivity. It seems that we have not learned to duplicate the conditions in the wild that allow proboscis monkeys to maintain healthy populations of symbiotic bacteria. Because we cannot save proboscis monkeys by preserving them in zoos, we must save their habitats. But the forests in which the monkeys live are some of the most endangered ecosystems on the planet.

How did the proboscis monkeys come to be painted into such a tight evolutionary corner? It is even tighter than that of the hoatzins, which can dine on a much wider range of plants. To see how this happened, we must look at the proboscis monkeys' relatives.

The primates, including ourselves, have radiated in many different directions from their remote insect-eating ancestors. Different members of the primate order have adapted to both generalist and specialist diets.

Proboscis monkeys are not the only primates that can ferment their food. All the other langur monkeys and their close relatives the colobus monkeys have a specialized fermentation chamber that has formed in front of their regular stomachs. These monkeys include the bewhiskered Hanuman langurs, like one that I encountered in central India's Bandhavgarh National

Figure 30 This Hanuman langur monkey in Bandhavgarh Park in India's central Madhya Pradesh State also carries out some fermentation of its food, but it can eat a far wider variety of foods than the proboscis monkey can.

Figure 31 Silverback Guhanda, the alpha male of the Sabyinyo group of mountain gorillas, is groomed by another male, Gukunda. These gorillas live in Rwanda's Volcanoes National Park, within sight of the Congolese border. Notice Guhanda's gently swelling stomach, a sign that he is fermenting his lunch of tender leaves and bamboo shoots.

Park—note its uncanny resemblance to Charles Darwin. But the langurs and colobus do not have as extreme a dietary specialization as the proboscis monkeys. Urbanized ones often dine on potato chips stolen from incautious tourists, without obvious ill effects.

New World howler monkeys and African gorillas share with the langurs and colobus monkeys the swollen bellies that are the external signs of fermentation, though leaves do not form such an overwhelming part of their diets and they do not have specialized foreguts. You can see the gentle fermentative swelling in the tummy of a blissful silverback gorilla that I watched being groomed by a younger male in Rwanda's Volcanoes National Park (see Figure 31).

The tiny galagos or bush babies of Africa, and several species of lemur in Madagascar, including the spectacular indri lemur, are also able to get some of their nourishment from vegetation with a high cellulose content. But these primates have expanded their mid-guts rather than their stomachs, and their fermentative bacteria reside in this mid-region of their digestive tracts. The swelling that results from this gut fermentation is not as obvious as it is in proboscis monkeys and the other langurs. Mid-gut fermenters are also able to eat a wider range of foods than the proboscis monkeys, because they do not depend as strongly on fermentation.

We ourselves are at a different dietary extreme. We non-fermenters are the most omnivorous of the present-day primates, though we share the title with the cute little capuchin monkeys (the "organ grinder" monkeys) of South America. These monkeys are also catholic in their tastes, dining with equal avidity on small mammals, insects, fruits, nuts, and tubers. No fermentation takes place in their stomachs or mid-guts.[8]

Capuchins have adapted to the destruction of their habitat by humans, because they are able to move into farmers' fields and eat crops like maize and sugar cane. And we, because of our own versatile digestive system, have been able to fan out across the entire planet. We are able to thrive on foods ranging from witchetty grubs to seal blubber. Indeed, this digestive versatility has played a huge role in our success as a species.

Evolutionary convergence towards the laid-back ruminant lifestyle

Now that the cast of characters is in place, we are ready to see how the evolution of ruminant-like digestion has taken the most probable of a number of possible paths.

Ruminant-like fermentation of foodstuffs has had a long evolutionary history. The ruminant lifestyle among browsing animals probably extends back into the Age of Dinosaurs. Well-preserved fossils of some of the early mammals that lived among the dinosaurs provide clues that they were probably fermenters. And a few equally well-preserved herbivorous dinosaur fossils show traces of enlarged stomachs that could also have been involved in fermentation.

Figure 32 The plaintive call of this Indri lemur, *Indri indri*, the largest of all Madagascar's lemurs, can be heard for miles in the island's eastern rainforests. Indri lemurs carry out fermentation in their mid-guts, rather than in their stomachs.

The fermenting lifestyle is so obviously advantageous that it is not surprising that fermentation of food has evolved more than once in different branches of the great tree of life. Termites, for example, are able to munch on our houses, not because they themselves can digest wood, but because their digestive tracts play host to thriving colonies of microorganisms. These organisms benefit the termites by breaking the wood down for them, but they eventually meet the same digestive fate that awaits the bacteria in mammalian ruminant guts. The termites acquired this symbiotic association quite independently of their remote relatives, the ruminants and other fermenting mammals.

Because fermentative symbioses have often evolved independently, it seems to have been relatively straightforward to adapt some region of one's gut in order to provide a place for thriving microbial populations. The advantage of this type of digestion is great when there is abundant food in the environment that is available in no other way. But we do not know the genes involved in these modifications of stomach anatomy.

We have, however, learned a great deal about one particular set of genes directly involved in fermentative digestion. In the 1980s the pioneering Berkeley molecular evolutionist Allan C. Wilson began studies that would eventually demonstrate how these genes have been co-opted to carry out the same task in similar ways in hoatzins and proboscis monkeys.

Wilson and his colleagues began by investigating the two main types of an enzyme called lysozyme that are made by animals.[9] All mammals, including ourselves, secrete the first type of lysozyme copiously in our tears and our saliva. It acts as an effective antibacterial agent, by breaking down the tough cell walls of any invading bacteria.

The second type of lysozyme is found in mammalian digestive systems. It is made in especially large amounts in the stomachs and foreguts of animals that live by fermentation. But this form of the enzyme is inactive when it is first secreted.

This initial inactivity is useful, because otherwise the enzyme would destroy the busily fermenting bacteria on which the animals depend. Then, as the mix of bacteria and food moves into the stomach's digestive chambers, the acidity rises and the change in pH activates the enzyme. The unsuspecting bacteria, swept from their carefree lives into the ruminant's digestive stomach, meet a swift fate as the newly activated lysozyme that is carried along with them chews through their cell walls and releases their eminently digestible contents.

Primates also have the gene for this stomach type of lysozyme. Although we humans and the other omnivorous primates make only small amounts of this lysozyme, the gene has been turned on full-blast in fermenting primates.

Wilson found that the stomach genes of monkeys and cows resemble each other, as do the monkey and cow tears-and-saliva genes.[10] But stomach genes from any mammal are distinctly different from tears-and-saliva genes. He concluded that an ancestor of the two types of gene had duplicated in the distant past, and the duplicates had begun to diverge to take up their separate functions even before the early divergence of the monkey and cow lineages.

The lysozyme in most birds behaves like the mammalian tears-and-saliva enzyme. The bird lysozyme is not secreted in the crop and takes no part in digestion. But it does make up about 3% of the whites of bird eggs. The egg lysozyme is active at the slightly alkaline pH typical of eggs and other tissues. This activity explains why eggs remain bacteria-free as the chicks develop. Bird egg lysozyme is a highly effective antibacterial agent. All bird lysozyme genes are extremely similar to each other and are quite distinct from the genes of mammalian lysozymes.

Hoatzin egg white lysozyme behaves like other bird lysozymes. But hoatzins have a total of five lysozyme genes, which arose through a process of gene duplication during the long period when the ancestors of hoatzins evolved on a separate lineage from other birds. Like the claws on the wings of hoatzin chicks, these duplicated genes have not been found in other living birds.

Three of these duplicated hoatzin genes have evolved away from the typical bird gene, and now code for the lysozymes that hoatzins secrete into their crops.

The forms of lysozyme secreted into the hoatzin crop behave differently from the normal bird lysozyme, and more like the mammalian stomach lysozyme. They are inactive under the typical alkaline conditions of the hoatzin crop, but become active under the acidic conditions found in the digestive stomach and in the gut.

One might expect that the crop lysozyme genes of the hoatzins and the stomach lysozyme genes of the fermenting mammals should have taken diverging evolutionary paths, even though their properties have now become similar. But Wilson and his colleagues made a remarkable discovery.[11] They found that, quite independently, the amino acids in the proteins coded by the hoatzin and langur genes had become identical at five different places. In effect, and independently, these two different proteins converged in their amino acid makeup at the same time as they converged on their new function of breaking down the cells walls of fermentative bacteria.

In view of all the possible changes that could have happened as the genes for this protein evolved in the hoatzin and langur evolutionary lineages, the likelihood that these ten independent events happened by chance and were not driven by natural selection is vanishingly small.

Allan Wilson tragically died at the age of 56 in 1991, at the height of his scientific career. But he was able to show in this path-breaking work that the genes of even quite different organisms like birds and monkeys can converge on similar chemical solutions to similar evolutionary problems. In the hoatzins and in the mammals, lysozyme genes have duplicated. The duplicated genes have evolved and converged to take up similar functions in fermentative digestion. Such a scenario is far more likely to happen than the appearance of a brand new gene that has lysozyme properties. Again and again, natural selection achieves its remarkable results by a process that Stephen Jay Gould called *exaptation*. Through exaptation, natural selection modifies genes that had originally evolved for different functions.

Competition and cooperation

There is much more to evolution than simple adaptation to new and changing environments. Natural selection can also produce cooperative behaviors between the individuals of a species, and between members of different species, across what might at first seem to be gigantic evolutionary chasms. The relationship between fermenting animals and the microbes that live in their guts is an excellent example of such cooperative

Figure 33 A yellow-billed oxpecker, *Buphagus africanus*, cleans ticks from the nostril of a relaxed cape buffalo in the Okavango Delta, Botswana.

bridge-building, leading to a kind of "super-organism" that has enhanced competitive ability.

Fermenting animals, whether they are ruminants or termites, cannot survive without their microbes. In many cases the microbes have in turn become so specialized that they cannot survive outside the stomachs of their hosts. But together the animals and their microbes have produced super-organisms that can efficiently exploit food sources denied to the rest of us. The microbes give their hosts the energy to flee from their predators, and in return they gain a safe place to live. Even though their individual lives are short, invariably terminating when their hosts digest them, their populations persist and thrive. The blanket term for such interactions is *symbiosis* or living together. The types of symbioses that provide mutual benefit for the different organisms involved are called *mutualisms*.

There are many other mutualistic interactions between species, though most of these interactions do not rise to the level of the total interdependence that exists between the grazing animals and their microbes. I recently watched as crowds of red-billed oxpecker birds strutted undisturbed over the faces and shoulders of a large herd of Cape buffalo in Botswana's Okavango delta. The birds were dining well on ticks and botfly larvae that had attached themselves to the buffalo, often enthusiastically thrusting their heads up their hosts' nostrils in the process.

The mutual benefit here was obvious. The buffalo get to live blessedly tick-free lives and the birds get to nibble on blood-filled tick canapés. The oxpeckers have even been observed to press a little blood from the animals' bites as they pull the insects free. But their buffalo hosts do not begrudge the birds this extra little *amuse-bouche*, because the relief that the oxpeckers provide is so palpable.

Oxpeckers perform their ministrations on many different species of grazing animal, enhancing the quality of their various hosts' lives while at the same time benefiting themselves. And they have influenced the behaviors of their hosts, who tolerate the most intimate ministrations of these cheery birds.

The oxpeckers and their hosts do not form a true super-organism. Much of the time the birds must live separate lives from the grazing animals. They must find mates, build nests, and feed their young, all the while being alert to a wide variety of predators. The grazing animals, too, live lives that are benefited by, but are largely separate from, the oxpeckers. The perils that they face from lions, leopards, and cheetahs are just as great as if the oxpeckers were not there.

There is a whole range of mutualistic interactions that lie along the spectrum between the kind of intimate association that exists between Cape buffalo and their fermenting microbes and the more casual but still friendly interactions between the buffalo and their oxpeckers.

It would be lovely if all organisms could cooperate, like the Cape buffalo and the oxpecker, or the hoatzin and its thriving microbes. Then the world might come to resemble American Edward Hicks' famous painting *The Peaceable Kingdom*, in which the lion lies down with the lamb. But cooperation, though it is often favored by natural selection because it produces a super-organism, can only extend so far. The super-organism must still compete for food and defend itself against danger. All cannot be sweetness and light in the biological world. Sooner or later, even super-organisms must eat or be eaten.

Turning people into hoatzins

We humans, unlike the proboscis monkeys, can easily adapt to new food-stuffs because of our evolutionary history as an omnivore. But there are limits to our own adaptation, at least in the short term. To illustrate this, let us explore an all-too-plausible near future in which our population has increased to the point at which we have irreversibly damaged our planet, reducing the amount of food on which we all depend. Could we save ourselves by taking advantage of cellulose, the immense food source that proboscis monkeys thrive on but that is completely unavailable to us? What kinds of evolutionary change would that new ability require?

Let us begin with the food potential of cellulose. As of this writing the Sunday edition of the *New York Times* has slimmed considerably, but before the newspaper business tanked and news junkies decamped to cyberspace this great mass of newsprint weighed about two kilograms. Much of the paper's bulk was made up of ads touting quaintly antediluvian mechanical wristwatches, at prices that could retire the national debt of a smallish country. But interspersed among these ads was all the news that's fit to print. If you lingered abed each Sunday and read the paper thoroughly, you would be sure to become almost pathologically well informed.

But then you had to get rid of that great mass of paper. At the moment the best way is to recycle it. A third of the four billion trees cut down worldwide each year are made into paper, and more efficient recycling could reduce this number substantially. Luckily, the habit of paper recycling is growing dramatically in the USA, and currently more than 50% of the paper in the country is salvaged and recycled. But the rest ends up in landfills, or as ash and carbon dioxide. And recycling is strongly dependent on the world economy, so that the financial payoff for recycling goes up when our economy is healthy and goes down when the economy is sick.

Suppose that, instead of recycling the paper, you could eat it? The *New York Times* is made of fibers of cellulose, and as we saw earlier the cellulose in turn is made up of long strings of the nourishing sugar glucose. Luckily for the newspaper's food potential, the process of paper-making has removed the indigestible hemicelluloses and lignins that make life difficult for the bacteria in the crops of hoatzins and the stomachs of proboscis monkeys.

If you could enzymatically unlock these strings of sugar into their component parts by digesting the paper, you would gain the energy contained in two kilograms of the sugar glucose, roughly 8,000 calories. At a liberal daily allowance of 2,000 calories a day you could live for four days on one Sunday paper! Of course you might prefer the sports or the food and dining sections to the glossy magazine. The magazine's pages have been covered with a truly indigestible layer of clay-based coating, in order to make the paper shiny.

Hoatzins could in theory dine on the *New York Times*, though they would certainly not care for the taste. So what would it take for us to do the same thing? If our close relatives the proboscis monkeys and the gorillas can utilize cellulose by fermenting it, what is to prevent us from solving our food problem by following in their footsteps? It may only be necessary for us to tweak a relatively small number of our genes.

The hoatzins' crops have become modified, enabling them to carry out long periods of fermentation. Their behaviors have also changed. To accommodate the fact that they are clumsy flyers they build their nests on the banks of rivers and lakes, so that they are less likely to crash into trees as they take off and land. And of course they have a strong preference for diets made up of leaves.

All these changes have been the result of natural selection. Could equivalent changes, suitably tailored to our own needs, be introduced into our own species?

Suppose we could modify our digestive systems so that they became more like those of the hoatzins or proboscis monkeys. Some of the evolutionary work has already been done. We already have a lysozyme gene that has the potential to work well in a fermentative stomach, because some of our ancestors had to adapt to a fermentative way of life repeatedly in the past. In omnivores like us this gene has been turned off, but it would be relatively simple to turn it on again.

More dramatically, we would need to modify the structure of our stomachs. The stomachs of fermenting primates such as the langurs and proboscis monkeys have swollen greatly in size compared with ours. They consist of four distinct chambers. We too have four regions in our stomachs: the fundus that nestles against our esophagus, followed by the main body, the antrum, and finally the pylorus that leads to the small intestine. But in us these regions are not anatomically distinct. They would need to be separated before our stomachs could function like those of the langurs.

Our first two chambers, the fundus and the main body, could in theory be increased in size and modified to secrete less acid and make them more bacteria-friendly. Then our antrum could be pinched and narrowed, to separate the fermentative from the digestive stomach, and the pylorus could be modified to digest the bacteria and the products of fermentation. In carrying out these modifications we would simply be recapitulating the evolution of the langur's complicated stomach from the single-chambered stomachs of its non-fermenting ancestors.

Of course we would have to resign ourselves to a change in lifestyle. We would have to come up with ways to make the *New York Times* palatable, even tempting, after we have finished reading it. This might be done through the use of spiced or exotically flavored inks. There could be jalapeño and nacho cheese editions. Our stomachs would also swell embarrassingly, and we would have to get used to frequent, conversation-halting, methane-bearing eructations. Books of etiquette would have to be amended accordingly.

But would these drastic physiological changes help with the planetary food shortage? Cellulose is certainly plentiful, but like our current food supply it is not unlimited. The world produces about 300 million metric tons of paper and paper products every year. If we could eat it all and ferment it without waste, we could support a population of about 6 billion people, roughly the world's current population, at about 2,000 calories a day.

But 100% efficiency is never achievable. If a still-respectable 50% of the cellulose in paper could be converted to food energy, this would only be enough to support half of the world's current population on an adequate diet. And of course our population is heading towards nine billion by the middle of the century, making a cellulose solution to the world's food problem even less practical.

I was rather surprised when I looked at these numbers. Our population is so gigantic that even if we were all retrofitted to be able to munch on newspapers and cardboard boxes every day, followed perhaps by a paperback romance novel for dessert, we would still be faced with widespread starvation unless we chopped down even more forests than we do now.

It would simply not pay us as an entire species to carry out these drastic modifications. But this new lifestyle might appeal to those of us who would like to move beyond a vegan diet to the ultimate in recycling. A swollen stomach and inflammable belches might become proud signs of a raised ecological consciousness.

Evolution and its limits

As we contemplate the difficulty of feeding our growing population, we confront the same biological limit that gave Charles Darwin his first clue to the mechanism of natural selection. When he returned to England from the voyage of the *Beagle*, Darwin read a gloomy book by the clergyman T. R. Malthus, called *On Population*. Malthus pointed out that our population increases geometrically, while the resources on which we depend increase linearly if at all.

Darwin realized that Malthus' limits on available resources apply to every species. His great insight was that as the resource limits are approached, some members of the population will have a better chance of surviving than others. It is this differential survival that results in natural selection.

Despite our recent population explosion we are not immune from these Malthusian limits. We are now six times as numerous as we were when Darwin returned to England on the *Beagle* in 1835. Even drastic genetic engineering of our digestive systems, turning us all into methane-belching fermenters, would not be enough to maintain our present and future numbers. We may be able to postpone the inevitable for a little while, but it is clear that we must either control our own population or have a much less forgiving Mother Nature do it for us.

This chapter has given you a brief sketch of how evolution actually takes place. We have seen how genetic variation, the tank of gas that powers evolutionary change, is replenished and increased through mutation and natural selection. We have also seen how we as a species are not immune from the evolutionary pressures that have shaped the living world. Now let us look at how our changing planet itself, working in conjunction with biological evolution, has influenced the story of life. We will explore a nexus of geological change and biological evolution that is taking place where two great regions of the Earth's crust collide.

3
The Shifting Earth

Figure 34 Visible on the far horizon, the small volcanically active island of Savo juts up from Ironbottom Sound in the Solomon Islands. Savo is home to megapode birds that tap the energy of the Earth's shifting crust in order to survive.

The morning of February 8, 2008, dawned without noticeable fanfare on Yap Island. This tiny speck of land in the mid-Pacific Ocean is part of a scattering of islands, spread over an area the size of the continental United States, that makes up the newly independent nation of Micronesia.

I greeted the day with cheerful anticipation, as I invariably do when I am about to plunge into a tropical sea. But I did not anticipate that before the day was done I would have a personal encounter with the immense forces that have built mighty stretches of mountains, opened up gigantic canyons in the ocean floor, and moved islands and continents around from one end of the Earth to the other.

Yap is famous for the enormous crudely shaped stone disks, each with a hole in its center, that serve as signatures of wealth for the Yapese people.¹ The disks, accumulated and arranged in rows near each village meeting house, serve as money. They are surely the clumsiest means of commerce ever invented. I cannot image how one makes change for a stone disk four meters in diameter!

American dollars are the preferred currency on the island at the moment, but the disks still play a role in ceremonial exchanges between villages. The more of them that a village possesses, the greater its prestige. The stones are not local—all of them were quarried from quartz-bearing limestone on the island of Babeldaob, some 200 kilometers to the southwest. Each disk has a value in part determined by the number of people who died bringing it by canoe across the open ocean to Yap.

Yap is also famous among scuba divers for its manta ray cleaning stations. These are sites on the surrounding reef where giant manta rays line up, like cars at a service station, to be cleaned of parasites by busy wrasses and other small fish.

For several days I had been trying to get some good pictures of this behavior, but had been frustrated by cloudy water. This was the last day I could dive on the island—I was flying out the next evening, and I had to leave a gap of 24 hours between my last dive and the airplane flight in order to clear any extra nitrogen from my body.

Figure 35 A meeting house on Yap, with stone money disks. The larger the disk, and the more lives that were lost bringing it to Yap, the greater its worth.

Nature's jackhammer

Water conditions were still not good in the afternoon, because an outgoing tide was carrying silt from the island out to sea. Nobody else on the island was diving. Nonetheless, divemaster Mike Kuiper, his dive buddy Diane Corbière, and Barney our Yapese dive guide were kind enough to head out with me to the Miil channel where the rays habitually cruise through a gap in the reef. We dropped to the sandy bottom, in about twenty meters of still distressingly murky water, and swam to some coral outcrops at the base of the channel's steep slope. There we waited for the mantas to appear.

Nothing much happened for the first half hour. Then a manta swam by, followed by another, and a third. Each was about three meters from wingtip to wingtip, and they ignored us as we kept close to the bottom and snapped pictures. Once again, even though the mantas swept by just above our heads, it was too cloudy to see much cleaning activity.

Then, just as I had taken a picture of a manta that was swimming past only three meters away, I was suddenly buffeted from behind by a series of physical blows accompanied by loud hammering noises. The blows were so powerful that I could feel them in my chest. They continued at full force for what seemed like an endless time but was probably only about thirty seconds. Then they gradually diminished. The manta that I had been photographing had, very sensibly, vanished.

In trying to describe the sensation afterwards, I could only compare it to an intimate encounter with a pneumatic drill that I had experienced decades earlier. While I was helping to demolish a building in order to pay my way through college, I was given the massive air hammer by my boss. He told me to tear up a concrete basement floor with it. The hammering I had just experienced underwater was as loud and violent as my youthful *pas de deux* with that pneumatic drill.

Figure 36 Brilliantly colored mandarin fish, *Synchiropus splendidus*, mate on a reef near Yap. The fish lurk in the corals and wait shyly until dusk to consummate their brief affairs.

As the vibrations died away we all stared at each other with astonishment and worry. What on earth had happened? My first thought was that a high-pressure fitting on one of our air tanks had ruptured. But this would have released a blast of air bubbles into the water. There were no telltale bursts of bubbles, either from me or from my fellow divers. We glanced at our dive computers, all of which gave perfectly normal readings.

Then I looked up and saw a huge cloud of mud and debris flowing in slow motion towards us down the slope of the channel.

It was now clear what had happened. We had been caught in a powerful earthquake that had triggered an underwater landslide. I signaled to the others to alert them to the oncoming debris, and we rose above the mud cloud and headed towards the surface.

We had not been very deep for very long, so we did not need to pause for decompression. But we did need to do a safety stop at three meters for three minutes, a period that after our experience seemed to stretch even longer than it usually did.

When we did break the surface everything seemed quite normal. But Eddie the skiff operator waved and shouted: "Hey! You guys tried to play tricks on me!" "What tricks?" we asked. He laughed and said, "You banged on the bottom of the boat!"

We assured Eddie that it was Mother Nature, not us, who had banged on the bottom of the boat.

Mike and Diane dove down again to inspect the damage to the reef. I had been through quite enough excitement for one day, and stayed on the surface.[2] They came up to report that they had felt two aftershocks, and that they had found a chunk of coral three meters in diameter that had broken off the reef near the surface and rolled down close to where we had been diving. And they saw mantas, which had calmly returned to the cleaning station as if nothing had happened. We realized that we now had direct proof that mantas show no obvious behavioral changes before an earthquake, unlike dogs and domestic animals who have often been reported to behave strangely. Either the mantas had not anticipated the earthquake, or they simply didn't care.

After these dramatic underwater events we headed back to Yap's largest town, Colonia, to find out how the people above water had fared.

Figure 37 A giant Manta ray, *Manta birostris*, in the Miil Channel off the island of Yap, seconds before the earthquake struck.

Extremely well, it turned out. Everybody in the town had felt the quake, but nothing had been damaged. A few old Japanese beer bottles in the dive shop had fallen over. My wife Liz had been in the hotel swimming pool when the quake hit. She felt a slight sloshing sensation and heard a sound like a truck going by.

Under the sea it was a different story. Over the next few weeks, after I had left the island, Mike and Diane dove in other places where they were familiar with the configuration of the reef from numerous previous dives. They found that many massive slabs of coral had broken free and slid down slopes on the reef.

Consulting the web, Mike discovered that the quake's epicenter had been only ten kilometers from where we had been diving. Its strength had been a hefty 5.4 on the Richter scale. The difference between the effects of the quake above and beneath the water was astonishing. We realized that the water, eight hundred times as dense as air, had magnified the pounding effect of the compression waves generated by the quake. We also realized that otherwise inexplicable features of coral reefs that we had encountered at many places over the years, such as masses of dead corals lying deep underwater where the sunlight could not reach them, and table corals that had been tipped away from the horizontal, must have been the result of earthquake activity.

I had experienced a number of minor earthquakes above water, but had never encountered the true power of our endlessly shifting planet until that moment in the Miil Channel. Mike told me that he had actually been under-water in the Maldive Islands during the 2004 tsunami that had swept across

Figure 38 This picture is the only one I had the presence of mind to take of the underwater landslide following the earthquake.

the width of the Indian Ocean. He felt the surge of the wave sweeping him towards the reef and then away again, but there was no damage to the reef itself.

The compression waves from that far stronger earthquake had weakened by the time they reached the Maldives, because that quake's epicenter had been off the coast of Sumatra, 2500 kilometers away. But the tsunami had caused a pattern of damage that was the mirror image of our Yap earthquake. When Mike returned to Malé, the Maldivian capital, he found that the wave, which had little effect on the surrounding reefs, had devastated the islands themselves. More than a hundred people had been killed and 20,000 made homeless as the islands were swept by the huge tsunami.

In this chapter we will see how a similar encounter with an earthquake helped Charles Darwin formulate his theory of evolution. Then we will explore the Earth's greatest current zone of earthquake and volcanic activity, to see how living things have reacted to geological upheavals large and small. We will discover that life is astonishingly resilient, and has managed to survive even the largest upheavals. And we will see how these upheavals have provided new opportunities for evolution.

Darwin's jackhammer

Our little group of divers on Yap had been given a fresh reminder of the immense forces that lie beneath our feet on this unstable planet. Charles Darwin had a similar introduction to these forces, one that preceded ours by almost two centuries. The huge forces that have shaped our world, and the immense spans of time over which they have acted, were brought home to Darwin during his voyage on the Beagle.[3]

By the beginning of 1835 the crew of the Beagle had spent more than a year mapping the complicated and dangerous channels and reefs of Cape Horn. Their surveys, which were the first to use chronometers to measure the precise longitude of these hazards, would save the lives of many voyagers. Now the Beagle was leaving this area of wind and storms to begin a long journey of exploration up the west coast of the South American continent.

The *Beagle* was exploring a world of astonishing geological activity, far more obvious in its effects than the processes that had shaped the English countryside with which Darwin was familiar. Indeed, just before embarking on the voyage, Darwin had taken a field trip to Wales with Cambridge's geology professor Adam Sedgwick. Both of them—as Darwin later admitted—had totally missed the signs of the glaciers that had carved and shaped the Welsh valleys.[4]

But nobody could miss the geological processes at work in Tierra del Fuego and the tip of Chile. When Darwin arrived at the bay of San Carlos in southern Chile in January 1835, he had already seen the dramatic effects of glaciers in shaping Chile's southern tip. Now he watched in wonder as the nearby volcano Osorno hurled fireballs into the sky. Later he learned that two other giant volcanoes, spaced along South America's western coast over a span of 4,000 kilometers, had been in full eruption at the same time.

As the *Beagle* sailed northward along South America's west coast, it was almost always within sight of the Andes, the world's second highest mountain range and one of its youngest. Darwin was aware of the argument between catastrophist and uniformitarian geologists over how such immense mountains had been built. Catastrophists assumed that such immense features had been produced by a single event such as Noah's flood, or by a series of catastrophes that marked periods of great upheaval in the Earth's history. But Darwin favored the theory of the geologist Charles Lyell, who built on earlier ideas of James Hutton and proposed that the Earth's geological features can be explained by the processes we see around us. Mountains, Lyell maintained, are built up gradually through volcanic activity and earthquakes, and broken down equally slowly through erosion. He called his theory of gradual change uniformitarianism. Darwin, confronted by mountains larger than he had ever seen, was determined to test Lyell's hypothesis.

In order to explain the slow formation of such immense geological features as the Andes, Lyell's uniformitarian theory postulated that the Earth was ancient, and that spans of time in the order of hundreds of millions of years must have passed. When Darwin later formulated his theory of evolution, he too made the assumption that the world has existed for great stretches of time. This would, he postulated, have given sufficient time for life's present-day diversity to evolve. But on the Chilean coast Darwin discovered that geological changes could be startlingly sudden—not necessarily the major catastrophes of the old-fashioned geologists, but mini-catastrophes that when summed together could accomplish major changes.

On February 20th, Darwin went ashore to collect specimens near the tiny Chilean town of Concepciòn. While he was resting in a little wood nearby the ground began to shake violently. It turned out to be the strongest earthquake that had been recorded in the region up to that time.

Even as the earthquake was happening Darwin's superb powers of observation did not desert him. He watched as the ground visibly undulated, and tried to determine from the waves the direction from which the quake had originated. As an experiment he attempted to stand up, and found himself swaying as if drunk. Afterwards he wrote in his journal: "A bad earthquake at once destroys our oldest associations: the earth, the very emblem of solidity, has moved beneath our feet like a thin crust over a fluid;—one second of time has created in the mind a strange sense of insecurity, which hours of reflection would not have produced."

Although the earthquake had lasted for two minutes, the forest that surrounded him seemed unaffected. But when he traveled to Concepciòn some days later he found the town in ruins. A hundred people along that part of the coast had been killed by collapsing buildings and by an eight-meter tidal wave. A schooner that had been anchored in Concepciòn's harbor was carried 200 meters inland.

Darwin's earthquake was far more powerful than the relatively mild earthquake I had experienced on the fringing reef of Yap. In my case water had magnified the earthquake's power, damaging the reef and battering me and my fellow divers but causing no damage on land. Darwin was astonished by the destructive effect of the compression waves that had been generated by the far larger Chilean earthquake. He noted, as he explored a nearby offshore island: "The effect of the vibration on the hard primary slate, which composes the foundation of the island, was…more curious: the superficial parts of some narrow ridges were as completely shivered as if they had been blasted by gunpowder." It is a good thing that scuba gear was not available in 1835, or Darwin might have been collecting specimens underwater, where he would almost certainly have been killed by those intense compression waves.

Members of the crew soon discovered even more dramatic effects of the earthquake. As they explored the coast by small boat they found that some of the mussels, barnacles, and other animals that normally lived in the intertidal zone along the shore were now rotting in the sun above the level of the highest tide. In a period of minutes the earthquake had raised kilometers of coastline more than two meters into the air.*

Darwin soon realized that this earthquake, violent as it was, must have been only a small contributor to the much larger pattern of events that had sculpted the geology of the entire western coast of South America.

Just inland from the Chilean coast the land slopes dramatically upward towards the Andes, the second highest mountain range in the world. A month after the earthquake, Darwin had the opportunity to traverse the mountains through a high pass. At 4,000 meters he observed beds of marine sediments that contained fossils of mollusk shells from species similar to those living in the deep sea today. It was clear that these fossil beds, and indeed the entire mountain range, had been lifted through the cumulative effect of many earthquakes, some of which had probably been far more violent than the one he had experienced:

All the main valleys in the Cordillera are characterized by having, on both sides, a fringe or terrace of shingle and sand, rudely stratified, and generally of considerable thickness...No one fact in the geology of South America, interested me more than these terraces of...shingle. They precisely resemble in composition the matter which the torrents in each valley would deposit, if they were checked in their course by any cause, such as entering a lake or arm of the sea[.]...If this be so, and I cannot doubt it, the grand and broken chain of the [Andean] Cordillera, instead of having been suddenly thrown up, as was till lately the universal, and still is the common opinion of geologists, has been slowly upheaved in mass, in the same gradual manner as the coasts of the Atlantic and Pacific have risen within the recent period.

* On February 27, 2010, a massive earthquake measuring 8.8 on the Richter Scale hit the Chilean coast. It was centered near Darwin's earthquake of almost exactly 175 years earlier, but the damage was far greater because Concepciòn is now Chile's second largest city. The inexorable forces that are pushing the Andes upward continue to endanger people who live along the Ring of Fire.

Figure 39 Ecuador's Cotopaxi volcano, which last erupted in 1975, is typical of the numerous volcanoes that mark the collision of the South American and Nazca tectonic plates. In 1534 a violent eruption of Cotapaxi brought a sudden end to a battle between Incas and Spaniards.

We now know, as Darwin did not, that the Andes are the result of a collision between two tectonic plates, regions of the Earth's crust that move slowly relative to each other. During the past 60 million years the Nazca plate that makes up part of the Pacific's floor has moved eastward, sliding gradually beneath the western edge of the South American plate and crumpling it upwards. Strata have piled up on top of each other, pushing what had once been seafloor sediment higher and higher into the air. The eventual result was the Andes.

Darwin's trip across the Andes shaped his thinking about the natural world in other important ways. He was amazed by the differences that he found between the animals and plants on the western and the eastern slopes of the mountain range. It was as if they had come from two separate worlds, with quite different animals and plants filling the same ecological niches on the two sides:

I was much struck with the marked difference between the vegetation of these eastern valleys and those on the Chilian side: yet the climate, as well as the kind of soil, is nearly the same, and the difference of longitude very trifling. The same remark holds good with the quadrupeds, and in a lesser degree with the birds and insects. I may instance the mice, of which I obtained thirteen species on the shores of the Atlantic, and five on the Pacific, and not one of them is identical.

Although Darwin knew nothing of the tectonic plate movements that had caused the earthquake, his direct encounter with the Earth's unstable crust and the many other geological observations that he made during the *Beagle*'s voyage reinforced what he had earlier absorbed from his intensive reading of Lyell's books. On his trip he gathered further proof that our planet has had a long history, far longer than the cramped chronology still insisted upon by believers in the literal interpretation of the Bible. He also realized that small geological events can have mighty consequences when they are repeated many times. And, central to his growing conviction that species were not immutable and had evolved, he saw at first hand how the gradual emergence of the geographical barrier of the Andes had led to the formation of different species on the mountains' eastern and western slopes.

The origins of Indonesia—a collision of worlds

My relatively minor earthquake experience and Darwin's more substantial one are the kinds of geological changes that many people encounter during their lifetimes. Summed over a span of hundreds of millions of years, such small, medium, and large earthquakes and eruptions have led to dramatic rearrangements of the crust of our planet. One of the most impressive of these rearrangements is a vast slow-motion collision at the junction of the Pacific and Indian Oceans, a geological maelstrom that has led to the forma-tion of Indonesia and the Philippines.

This collision is bringing together animals and plants that have followed separate evolutionary paths for two hundred million years. In this and in subsequent chapters we will follow the consequences of that collision.

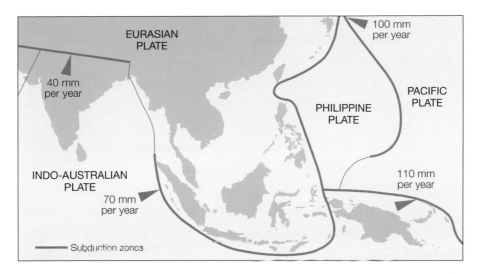

Figure 40 The current configuration of Indonesia's tectonic plates, with the location of the 2004 earthquake.

By the end of the nineteenth century evidence was accumulating from geology and the fossil record that the Earth's continents have not always been in their present positions. For example, nineteenth-century geologists found fossils of *Glossopteris*, a little fern-like seed plant, in ancient rocks in widely scattered locations in South America, Africa, India, and Australia. Much later, in 1965, *Glossopteris* was discovered to have thrived in Antarctica.[5] This and other fossil finds suggested that all the southern continents had once been joined together into a single vast landmass, later named Gondwana. We now know that Gondwana did exist, and that it began to break up 250 million years ago.

Many such otherwise inexplicable observations were brought together by the German geologist and explorer Alfred Wegener in 1915. Wegener concluded that the continents must have drifted somehow to their present positions, but his theory of continental drift was derided by geologists who pointed out that there was no obvious mechanism by which such gigantic masses of rock might have moved.

In 1927 English geologist Arthur Holmes suggested how the continents might have performed their trick. Holmes proposed a dynamic theory of tectonic plates (tectonic comes from the Greek for "builder"). He suggested that the Earth's crust, including the ocean bottoms, is broken up into great plates that float on the molten magma that lies below them. Convective cells in the magma provide the motive force for the plates.[6]

In the 1960s the discovery of traces of residual magnetism in old rocks showed Holmes' theory to be partly correct. The discovery relied on the fact that every few hundred thousand years the Earth's magnetic field switches poles, so that the north magnetic pole becomes the south and vice versa.

When lava is vented and solidifies, it tends to preserve the way in which the magnetic field was oriented at the time of the eruption. And it was the pattern in which this residual magnetism is preserved in the solidified lava of the mid-Atlantic ridge and other great underwater mountain ridges in the ocean basins that provided evidence for Holmes' tectonic plate theory.

The midline of each of these oceanic ridges turns out to be flanked by alternating bands of residual magnetism with switched orientations. The striped pattern of these alternating bands forms a perfect mirror image on each side of the ridge. And the widths of the various bands themselves reflect perfectly the various times at which the Earth's magnetic field has switched in the past. This alternating pattern could only be explained if the ocean bottom on each side of the mid-ocean ridge is being continually spread apart as fresh flows of magma at the ridge accumulate.

Continuing pressure from the spreading ridges helps to push tectonic plates apart as they float on the liquid magma. The spreading plates bump up against the margins of other plates and begin to descend beneath them. Much of the motive force that moves the plates comes from these descending plate margins, which pull the plates across the magma, rather than from the convective cells that Holmes originally envisioned. These collisions can trigger earthquakes and eruptions far from the original source of pressure itself, such as the collision between the Nazca and the South American plate that threw Darwin off his feet. Nowhere are the consequences of these collisions more dramatic than in the world's most tectonically active region, Indonesia.

Just north of the Indonesian island of Sumatra, the Indo-Australian plate and the immense Eurasian plate grind up against each other along a ridge of

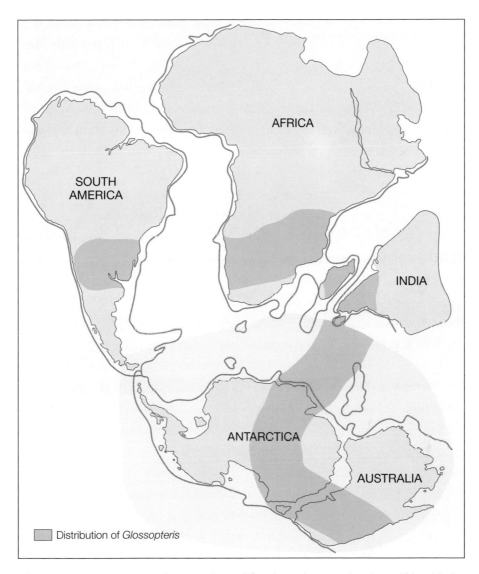

Distribution of *Glossopteris*

Figure 41   As the great southern continent of Gondwana began to break up, Africa, Madagascar and India moved northwards. The green areas show where fossils of the little early
seed plant *Glossopteris* have been found.

ocean floor. On Boxing Day 2004, the tension between the two plates was suddenly released. A 1600-kilometer stretch of the Indo-Australian plate slipped to the north, resulting in a powerful 9.2 earthquake that released half a million times as much energy as the little earthquake I encountered on Yap. As a result of the earthquake the Indo-Australian plate moved about twelve meters to the northeast. Different parts of the island of Sumatra rose and fell by two meters or more, shifts comparable in magnitude to the rise of the Chilean coast that the crew of the *Beagle* observed.

Another result of the released tension was that a region of seafloor along the slip zone suddenly rose several meters, pushing 30 cubic kilometers of ocean water upwards. It was this displacement that sent a tidal wave across the entire Indian Ocean basin, killing almost a quarter of a million people.

Indonesia is a region of island arcs, volcanoes, and deep-sea trenches that spans an area the size of the continental United States. It is daunting to think about how many such earthquake-triggered tectonic shifts, including many much larger ones that must have taken place before historical records began, have played a role in the formation of Indonesia's more than 17,000 islands.

The events that would lead to Indonesia's formation began 250 million years ago, at a time when the age of dinosaurs still lay 45 million years in the future. At that time the world's landmasses were concentrated into two great supercontinents. These in turn were the descendents of an even vaster supercontinent called Pangaea.

The northern continent of Laurasia would give rise to North America and Eurasia. The southern continent of Gondwana, which included land that would contribute to Africa, South America, Antarctica, and Australia, lay far to the west of Indonesia's present position.

It was at this time, a quarter of a billion years ago, that a severe mass extinction, the Permo-Triassic event, devastated both supercontinents. Ninety-five percent of the world's species were wiped out, apparently by vast volcanic eruptions. Among the many species lost was the little early seed plant *Glossopteris* that had thrived all over Gondwana.

Soon after the extinction event Gondwana began to break up. Its descendant fragments would eventually give rise to what are now Africa, India, South America, Antarctica, Australia, and New Zealand.

The first chunks to split off were Africa and India, calving off towards the northeast. Showing an early independent streak, India promptly began a surprisingly swift journey up the entire length of the nascent Indian ocean towards Laurasia.

The gap between Africa and the rest of Gondwana widened, accompanied by seafloor spreading in an arm of the ocean that would eventually become the Atlantic. The remainder of Gondwana, a complex region made up of nascent South America, Antarctica, and Australia, was still a single landmass, but it too began to break up. At the same time this whole chain of nascent continents moved south, so that the center of it, which would become Antarctica, entered the polar region. Antarctica remained warm, because the whole planet was warm, but its nights got longer and longer. Big-eyed nocturnal dinosaurs evolved in Antarctica.

South America finally broke away completely from Antarctica and Australia about 70 million years ago, near the beginning of the Age of Mammals. It entered a long period of isolation as a gigantic island. Australia and Antarctica then reached an irrevocable parting of the ways roughly 35 million years ago.

It was at about the time that Gondwana began to break up that three great lineages of mammals emerged from mammal-like ancestors called therapsids.[7] The most primitive, the egg-laying Monotremes, have left few fossils, and have survived down to the present only in Australia and New Guinea. This entire order of mammals now survives only as duck-billed platypuses and two species of echidnas—snuffly, Hoover-like spiny anteaters.

The second lineage, the Marsupials, raise their young in pouches. At the start of its life the little marsupial embryo is nourished briefly by a yolk-like placenta inside its mother's womb, but soon crawls out of the womb and attaches itself to a nipple in the pouch. Marsupial reproduction is adapted to desert life—under unfavorable conditions some kangaroos can evert their pouches and spontaneously abort the tiny joeys. This leaves the mothers free to start a new family when food is more plentiful.

The third group, the placental mammals including ourselves, have evolved a placenta, a mass of vascularized tissue that permits transfer of nutrients between mother and baby. The baby's development takes place entirely inside the mother's protected womb.

Figure 42 A wallaroo, *Macropus robustus*, leaps up a slope on Uluru (Ayers Rock) in central Australia. The ancestors of these pouched animals migrated early to the part of Gondwana that would become Australia and New Guinea.

Unfortunately, the fossil record tells us distressingly little about how mammals sorted themselves out in South America, Antarctica, and Australia during the critical period when these plates began to move apart. South America was home to thriving and diverse populations of marsupial and placental mammals, but Australia somehow ended up only with marsupials. Marsupials comprise most of Australia's mammal fossil record as well (though argument continues about some possible placental mammal finds in Queensland). Did the marsupials simply get to Oz first, and did Antarctica, growing as dark and dreary as Mordor as it continued to move south, act as a barrier to subsequent placental mammal invasions? Or did placental mammals lose out to better-adapted marsupials in Australia's tough environment?

My guess is that the marsupials were simply lucky enough to get to Australia first. They may have had a tough time doing so, because Australia shows signs of having been isolated even before the Age of Mammals. This, the furthest frontier of Gondwanaland, evolved its own unique collection of dinosaurs and, probably starting at about the same time, its own unique collection of marsupial mammals.

Australia eventually became home to most of the world's remaining marsupials, a few of which survived the dinosaur extinction and radiated once more. The resulting adaptive radiation produced marsupials that ranged from small rodent-like and squirrel-like animals through the familiar kangaroos, wallaroos, and wallabies to creatures such as the now-extinct Tasmanian wolf and other large and frighteningly toothy carnivores. One of these predators, the marsupial lion, was as heavy as present-day lions. It used its fused bladelike premolar teeth like giant shears, giving it a more powerful bite than any other animal that we know of, living or extinct.

The last of the catlike marsupials, the Tasmanian tiger, was driven to extinction in 1936, the same year in which Tasmania's farsighted government thoughtfully added the tiger's name to its protected wildlife list.

A change of course

Antarctica, straddling the South Pole, was now left behind as the Australasian plate that carried Australia began to move slowly north and east. As it moved it acted like a snowplow, pushing up at its northern margin the great cordillera of mountains that would eventually become the backbone of New Guinea. The plate's movement also helped to push up a series of island groups in the central Pacific. This chain of islands, which includes New Britain and the Solomon Islands, arcs out to the east of New Guinea. It marks the collision between the Australasian and Pacific plates.

As the great Australasian tectonic plate crunched further north it also began to encounter the eastern part of Laurasia, the great northern continent that had meanwhile broken up into North America and Eurasia. The Australasian plate began to thrust against the Southeast Asian portion of the Eurasian plate, also encountering the Philippine plate and other smaller chunks of crust. These smaller plates were pushed like bumper cars in many

Figure 43 Mount Inerie on the Indonesian island of Flores is typical of the volcanic land-scape throughout the Lesser Sunda Islands. Active and inactive volcanoes and cinder cones are everywhere, dotting the archipelago as if some race of gigantic moles has been at work.

different directions. All this activity helped to create the lesser Sunda chain of islands that stretches to the west of New Guinea, including Timor, Flores, and Sumbawa. At present the entire region between the Australasian and Southeast Asian plates has dissolved into geological anarchy. It is studded with seventy-six active volcanoes and shaken by hundreds of substantial earthquakes each year.

Some of the most complicated of these plate collisions formed the con-torted island of Sulawesi that we explored in Chapter 1. The other greater Sunda islands of Sumatra, Borneo, and Java, however, are not part of the Australasian plate. Instead they are upthrusted regions of the southeastern-most part of the Eurasian plate, pushed up by the tectonic collisions taking place to their south. These giant islands, surrounded by an expanse of shal-low seafloor known as the Sunda Shelf, share their geological and biological history with the great peninsula of Southeast Asia.

At the present time the Eurasian plate and the Australasian plate have not quite met. They are still separated by a complex of smaller plates. An arm of the ocean still forms a boundary between Borneo on the Southeast Asian side to the west and Sulawesi on the Australasian side to the east. But this arm is zipping up, starting from the south. Bali and Lombok, just south of Borneo and Sulawesi, used to be separated by a great southern extension of the oceanic arm, but this has now closed up and the islands are only separated by a narrow and relatively shallow strait through which strong currents pour from north to south.

Worlds in collision

Many of the animals and plants living on these two colliding plates have followed different evolutionary paths since Gondwana parted from Laurasia a quarter of a billion years ago. Now the gap between these two great biological provinces is a mere 25 kilometers wide at its narrowest point.

The birds, animals, and plants that rafted toward Southeast Asia on the Australasian plate were easily able to spread and colonize the new islands that had been thrust up to the north and east of New Guinea. These new islands, arising smoking from the sea, were empty of terrestrial life before the Australasian animals and plants arrived. But the Southeast Asian part of the vast Eurasian plate that lay to the west was already densely populated with well-adapted creatures. As a result, neither the Australasian nor the Southeast Asian biological province has been able to replace the other.

The pioneering biogeographer Alfred Russel Wallace was the first to document the existence of a line of demarcation between the two provinces. The discontinuity that he observed in the Malay islands was even more dramatic than the differences Darwin saw between the animals and plants on the west and east sides of the Andes, because its origins lay much further in the past. The abruptness of what became known as "Wallace's line" puzzled Wallace more than Darwin's Andean separation, because the geological reason for it is so much less obvious. The Southeast Asian and Indo-Australian biological provinces are divided, not by a towering mountain range, but by a far less intimidating narrow passage of ocean water between islands. Wallace could not have imagined the true reason for these differences, the long history of volcanic eruptions and tectonic shifts that has produced this biological Maginot Line.

The world knows Alfred Russel Wallace for his proposal, independently of Darwin, that evolution has taken place through natural selection—though of course it was Darwin who originated that memorable term. But there were in fact many differences between their views of evolution. One important difference was that Wallace emphasized that selection would result in the survival or extinction of entire races of animals or plants, while Darwin emphasized the accumulating effects of natural selection on populations of individuals.

Like Darwin, Wallace spent years piling up information about the natural world before he had his great insight. From 1854 to 1862 he explored the Malay Archipelago, collecting animals and plants and—unlike the independently wealthy Darwin—selling many of them to support his travels.

It was immediately clear to Wallace, when he crossed from Borneo or Bali in the west to Sulawesi or Lombok to the east, that the animals, plants, and birds changed utterly. His describes the transition vividly in *The Malay Archipelago* (1869):

The great contrast between the two divisions of the Archipelago is nowhere so abruptly exhibited as on passing from the island of Bali to that of Lombock, where the two regions are in closest proximity. In Bali we have barbets, fruit-thrushes, and woodpeckers; on passing over to Lombock these are seen no more, but we have abundance of cockatoos, honeysuckers, and brush-turkeys, which are equally unknown in Bali, or any island further west. [I was informed, however, that there were a few cockatoos at one spot on the west of Bali, showing that the intermingling of the productions of these islands is now going on.] The strait is here fifteen miles wide, so that we may pass in two hours from one great division of the earth to another, differing as essentially in their animal life as Europe does from America. If we travel from Java or Borneo to [Sulawesi] or the Moluccas, the difference is still more striking. In the first, the forests abound in monkeys of many kinds, wild cats, deer, civets, and otters, and numerous varieties of squirrels are constantly met with. In the latter none of these occur; but the prehensile-tailed Cuscus is almost the only terrestrial mammal seen, except wild pigs, which are found in all the islands, and deer (which have probably been recently introduced) in [Sulawesi] and the Moluccas. The birds which are most abundant in the Western Islands are woodpeckers, barbets, trogons, fruit-thrushes, and leaf-thrushes; they are seen daily, and form the great ornithological features of the country. In the Eastern Islands these are absolutely unknown, honeysuckers and small lories being the most common birds, so that the naturalist feels himself in a new world, and can hardly realize that he has passed from the one region to the other in a few days, without ever being out of sight of land.

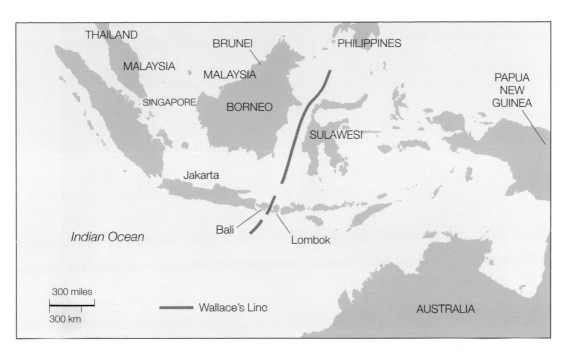

Figure 44 Wallace's line divides Indonesia. It marks the collision point between two great biological provinces — the southeast Asian province to the west and the Australasian province to the east

Sulawesi is the closest of the Indo-Australian islands to Wallace's line, which remains as abrupt as it was when Wallace discovered it. On a single afternoon in Tangkoko Park in northern Sulawesi, I encountered a placental mammal that had crossed Wallace's line towards the east and a marsupial that was the westernmost representative of the marsupials from Australasia. They live within a few miles of each other.

Visitors to Tangkoko spend much time trying to see another placental escapee across the line, the spectral tarsier. These tiny nocturnal prosimian primates emerge from the interstices of fig tree trunks just at dusk, to catch crickets and other insects in their little apelike hands. But the park, which protects a small part of the 10% of Sulawesi's forests that have not yet been destroyed, is full of other remarkable animals.

In an open forested area of the park I approached the bank of a stream and found myself surrounded by a rollicking and totally unafraid troop of

Figure 45 The Sulawesi spectral tarsier, *Tarsius tarsier*. A member of the Order Primates, to which we also belong, the tiny nocturnal tarsier has opposable thumbs and the largest eyes for its body size of any mammal. There are four subspecies of this placental animal on Sulawesi, all descended from a few relatively recent immigrants across Wallace's Line.

crested black macaques. These primates, too, have made it across Wallace's Line, perhaps like the tarsiers carried by trees uprooted in storms.

The younger macaques leaped from overhanging trees into the stream and played tag in the shallows, scattering great clouds of spray. They frolicked under the firm yellow gaze of the troop's patriarch.

These macaques are among the most prominent of the handful of placental animals, mostly small insectivores and bats, that have managed to

Figure 46 A male crested black macaque, *Macaca nigra*, another placental escapee across Wallace's Line, intently watches his playful troop in Tangkoko Park. This species of macaque is endemic to Sulawesi.

move east across Wallace's line to Sulawesi and its nearby islands. They have been on Sulawesi long enough to have evolved into a distinct species.

Not far from the macaque's forest, in an open grassland dotted with large trees, we caught a glimpse of something large and brown at the top of a huge chinaberry tree. It looked at first like a monkey, but as we threaded our way through the thick brush and approached the tree it was clear that it was a bear cuscus, the larger of two marsupial cuscus species found on the island. It gained its common name because of its rather endearing bearlike face. It clung to a branch with its prehensile tail and munched on young chinaberry leaves.

In this blurring of Wallace's Line we can see echoes of an earlier evolutionary battle between placental and marsupial animals. Two million years ago, placental mammals from North America were able to invade the previously isolated South America along the newly formed Isthmus of Panama.[8] As the placental mammals burst into South America they contrib-

Figure 47 This bear cuscus, *Ailurops ursinus*, also from Tangkoko Park, is one of the marsupial pouched animals that have penetrated furthest to the west on tectonic fragments being pushed by the great Australasian plate. But it has not crossed Wallace's Line. Scattered rumors of cuscus sightings on Borneo and Bali on the western side of the line seem to be unfounded.

uted to a wave of extinctions that wiped out almost all of South America's marsupials. Some marsupials managed to travel the other way, but most of these escapees eventually went extinct. Only the tough and versatile opossum survived.

At Wallace's boundary, too, placental mammals have been more successful in moving across the line to the east than marsupials have been in moving to the west. And humans have accelerated this invasion of the placentals, by introducing deer and monkey species to Lombok and Sulawesi.

The chaotic geology of Indonesia has resulted in an encounter between highly divergent groups of organisms, one that would not have happened if Australia and New Guinea had kept moving to the east instead of veering to the north. Life in this part of the world has survived literally thousands of geological catastrophes—vast volcanic eruptions, immense tidal waves, and the crunching descent of great pieces of the Earth's crust into the depths of the oceans. How, through all this terrifying cacophony, have living things managed to be so resilient?

Some of them did so by adapting to such extreme conditions. The story of the megapodes provides one example of life's infinite resourcefulness.

Danger and opportunity

Australasia's uniqueness is not confined to its marsupials and monotremes. Many of the birds of Australasia, such as the bowerbirds, birds of paradise, and megapodes, are also found nowhere else on the planet.

Like the hoatzins, the megapodes occupy their own distinct evolutionary lineage, though they are remotely related to chickens. They are also the only birds that have taken advantage of the volcanic activity in the Pacific's great ring of fire to help them reproduce. Megapodes are a poster child for life's astonishing ability to take advantage of potential disasters.

Megapodes bury their eggs where natural processes like organic decay or volcanic heat will keep them warm. The parents then wander off to live carefree megapode lives, while their chicks develop underground. After hatching the chicks actively dig their way to the surface. From the moment that they emerge they are the most precocious of baby birds, as expert at foraging as a month-old chicken. Because of this expertise they are immediately able to take up an independent life, in many cases without ever encountering their parents.

Megapodes ("big feet") are relatively common in Australia, where they are known as brush turkeys. Some of the megapodes are good flyers, which has enabled them to spread far to the north and colonize island groups like the Solomons. Pigeon-sized megapodes have even made it to middle-ocean islands like Palau and the Mariana Islands, far from their center of origin.

The eggs of megapodes are much-sought and delicious. (They have a richer and more intensely "eggy" flavor than chicken eggs—don't embarrass me by asking how I know.) As a result of uncontrolled egg predation, megapodes now only tend to survive far from human activity.

One of these spots is the tiny, mildly active volcano of Savo in the Solomon Islands. This heavily forested islet marks the entrance to a vast protected anchorage called Ironbottom Sound that lies to the north of Guadalcanal, the main island in the group. All the islands that surround and shelter the sound were pushed up by the collision of the Indo-Australian and Pacific tectonic plates.

Ironbottom Sound was once called Sealark Sound. It got its new name as a result of the Battle of Guadalcanal in 1942–3, which littered the sound's floor with the remains of more than fifty American and Japanese ships.

Recently I ventured across the sound to Savo in a little "tinny," a flat-bottomed aluminum boat powered by a decrepit outboard motor.

The people of Savo, like most of the Solomon Islanders, are Melanesians. They live off the crops that can be grown in abundance on the island's rich volcanic soil. Their diet of manioc, fruit, vegetables, and fish is varied by wild megapode eggs.

The villagers' diet is healthy, but they live in a dangerous world. As we climbed up to see the megapode nests, the headman of the little village of Reko told me (with a kind of perverse pride) that his villagers all suffer from repeated bouts of three out of the four different kinds of human malaria.

On the forested outer slopes of the Savo volcano the ground radiated a gentle warmth that I was able to feel through my boots. Hot rock lies so close

Figure 48 A piping hot waterfall on the slopes of the Savo volcano, Solomon Islands.

to the surface that it heats the streams that flow down the volcano's flanks. Scalding waterfalls nourish dense gardens of ferns and mosses. As I scrambled up the volcanic slopes through the hot mist I managed to catch glimpses of the megapodes, and saw the disturbed ashy soil that marked their nests. The villagers dug into one of the nests and found a brown egg, which they brought carefully back with them.

The Savo megapodes have evolved to take advantage of the energy generated by the same great volcanic upheavals that move continental plates around. They are able to excavate their nests in soft soil that is kept at a cozy temperature by the volcanic heat from below.

How long will the megapode population of Savo persist? Not long, probably. The villagers watch over the nests and are careful not to take too many eggs, but poachers sometimes sneak in and grab eggs clandestinely. Several species of megapode in the islands of Melanesia have already gone extinct.

Figure 49 The headman of Reko village on Savo digs out a megapode egg from the warm soil on the slopes of the volcano.

Figure 50 The megapodes on Savo eluded my camera, but I did catch this orange-footed scrubfowl, *Megapodius reinwardt*, on the Lesser Sunda Island of Rinca. It showed off its eponymous big feet as it pawed through leaves. These megapodes, denied the constant heat source that is available to the birds on Savo, pile up dead leaves and sticks to make their nests and depend on organic decay to keep their eggs warm.

It will be especially sad if these birds disappear, for they demonstrate how life can adapt to a dangerous world. Megapodes by the million have been roasted by sudden explosions of the volcanoes on which they depend for warmth. But some megapode populations have always managed to survive those localized disasters. Now, the new and far more widespread environmental changes that have been brought about by our own species may prove to be too much for them.

Extinctions and catastrophes

What effects have these repeated geological catastrophes had on the living world? We might expect that many of the huge eruptions and tsunamis would lead to waves of extinctions, but in fact most of them seem to have had surprisingly little effect.

Indonesia has had more than its fair share of such disasters. In 1883 the island of Krakatoa off the coast of Sumatra exploded, killing 35,000 people and producing a tsunami that traveled around the world at least twice. Earlier, in 1815, the 4,000-meter volcano of Tambora on the lesser Sunda island of Sumbawa erupted more violently than any other volcano in recorded history.

Tambora's explosion killed at least 70,000 people and left a crater 6 kilometers wide and 1,000 meters deep. The volcano blasted 150 cubic kilometers of dust, ash, and hydrogen sulfide into the air, forming a vast column 25 kilometers high that reached into the stratosphere.

Much of the ash probably settled quickly, but droplets of sulfuric acid were carried by stratospheric winds around the planet. The droplets reflected so much sunlight that temperatures fell, crops failed, and livestock died in Europe and America, causing widespread starvation. In New England the year 1816 was called "Eighteen hundred and froze to death," the year without a summer.

By comparison, the eruption of Oregon's Mount St. Helens in 1980 threw a trifling single cubic kilometer of ash into the air. Even though this amount of ash would have filled a hundred million large dump trucks, the eruption had little effect on climate, lowering temperatures in the northern hemisphere by about 0.1° Celsius.

But these eruptions, dramatic as they were, pale into insignificance as we reach further back in time. Indonesia has been the scene, not only of the most powerful eruption in recorded history, but of the most powerful we know about in the past two million years. Seventy-four thousand years ago an explosion far more powerful than Tambora's wiped out an entire mountain on the island of Sumatra.

All that now remains of the mountain is Lake Toba, a substantial body of water filling a crater that measures 100 by 30 kilometers. The calm waters of the lake are now plied by boatloads of Indonesian vacationers, most of them quite unaware of the violent origins of their pleasant holiday spot.

The mountain that disappeared has been posthumously given the name Mount Toba. Its explosion was indeed in a class by itself. It threw 3,000 cubic kilometers of ash into the air in a great tree-like plume that towered 80 kilometers to the fringes of outer space. This immense plume blew away to the northwest, covering much of what is now Malaysia with a smothering layer of ash and sifting down as far away as central India to form an ash layer that is 6 meters thick in places. It is estimated that for a decade afterward world temperatures were as low as they were during the peak of the last ice age.

Volcanic explosions such as Mount Toba leave many traces behind, but the effects of past tsunamis are more difficult to detect. We do know that tsunamis happen all the time—24 of them have hit North America over the past quarter of a millennium. And at least one tsunami has changed the course of civilization. When the caldera of the Mediterranean island of Santorini collapsed in about 1650 BCE, it produced a wave 100 meters high that swept away much of the thriving and highly advanced civilization of Crete, 150 kilometers to the south.

Many tsunamis of the distant past must have been even more violent, though the further back one goes in time the more uncertain the evidence becomes. It is generally agreed, however, that about 250,000 years ago an immense tsunami hit the islands of Hawaii. The wave was so huge that it deposited large rocks 250 meters above the sea level of that time.

What was the biological effect of these various disasters? Probably surprisingly little. Small events, like the one I experienced off the island of Yap, generally have little effect. The coral reefs near Yap that were damaged will soon heal. Fresh coral growth quickly knits together the rubble from old earthquakes and builds new living structures that reach toward the light.

Darwin's earthquake also had little effect on the non-human world. When he picked himself up off the ground after the Chilean earthquake, Darwin was struck by the fact that the forest around him seemed quite unperturbed. Even the millions of mussels and other intertidal organisms that the *Beagle*'s crew later found rotting above the high tide mark along the nearby coast were soon replaced. There are always clouds of eager larvae swarming in that cold, rich ocean, searching for empty places to settle down.

But what about the really big events? Surely the Toba catastrophe, for example, must have had an effect on us and on other animals. Stanley

Ambrose of the University of Illinois has proposed that the effects of Toba were so devastating that our own species nearly went extinct.[9]

Violent as the event was, most scientists are pretty doubtful that Toba had as severe an impact on our species as Ambrose suggests. To begin with, its effects were not planet-wide. Arctic ice cores show a trace of the Toba eruption in the form of a distinct pulse of sulfate. But a similar spike of sulfate has not been found in the Antarctic.

A few large southeast Asian mammal species might have disappeared after Toba, but they also might have disappeared because of general climate changes, human hunting, or simply the incompleteness of the fossil record. As for human activity, there is only one piece of evidence with a direct bearing on the question. Michael Petraglia of the University of Cambridge has shown that the stone tools made by early modern humans living in southern India at that time were the same before and after the eruption, even though these peoples were directly in the path of the ash fall.[10]

Intriguingly, there is genetic evidence that human populations outside Africa may have become reduced in size around the time of the Toba eruption, and Ambrose has suggested that the eruption was the cause. But there are large uncertainties about exactly when this reduction in population size took place, and much of the evidence has more than one interpretation.

We simply do not have enough data to measure the effect of Toba on our population in the past. Did the eruption nearly kill us off? Possibly, but if it was so disastrous then it seems strange that the eruption did not do more detectable damage to other animal and plant populations. In particular, it did not kill off the orangutans of Sumatra, even though they lived at the epicenter of the explosion.

The true evolutionary role of the Earth's upheavals

Though I did not enjoy my little earthquake at the time it happened, in retrospect I am pleased to have re-lived, in some small way, Darwin's experience with the Chilean earthquake. We both had the same kind of encounter with the massive forces that have shaped our Earth. My encounter with the power of earthquakes has made the true nature of Earth's geological history far more real to me.

As the Earth's tectonic plates slide and crash into each other in exquisite slow motion, they produce many local disasters that kill enormous numbers of individual animals and plants. Nonetheless, most of the volcanic eruptions and tidal waves have not wiped out whole species. Commonly, extinctions happen when a species becomes adapted to a narrow ecological niche and then the niche disappears. Before humans, such niches usually disappeared because of climate change or the invasion of new species.

If a species inhabits a broad range of niches and is spread over a large geographic range, then it is unlikely to go extinct. We humans, I suspect, were already occupying so many ecological niches in different parts of the Old World at the time of the Toba event that we could easily have survived an ordinary volcanic eruption, no matter how spectacular.

Even if my earthquake had somehow killed all the manta rays of Yap, other populations of this widely distributed species would not have been affected. A few days after the earthquake I dove on a manta cleaning station on the island of Pohnpei, 250 kilometers to the east of Yap. The mantas of this thriving population were being cleaned by wrasses that darted in and out of their gill slits. It is unlikely that even a force 9 earthquake on Yap would have affected the Pohnpei population.

Extinctions are also common when two tectonic plates approach each other for the first time and a flood of new animals and plants pour into a region. The invaders may ruthlessly take over the niches of the previous inhabitants. Such encounters happened when North and South America joined together through the Isthmus of Panama between three and two million years ago, ending South America's long isolation. A similar encounter is taking place in the Indo-Pacific as the Australasian and Southeast Asian plates continue to approach each other. Eventually, even in the absence of humans, Wallace's Line would have been breached and a flood of placental animals would have reached Australia.

Humans, of course, have accelerated the process. The invasion of Australia by Europeans and their domestic and wild animals has triggered repeated ecological disasters. Grazing and urbanization, along with the depredations of rabbits, feral domestic cats, and red foxes have driven about 10% of Australia's native marsupial species to extinction. Many more native species are threatened. Australia is home to about 10% of the world's endangered species, even though it makes up only 5% of the planet's land area and much of it is harsh desert.

Truly titanic events

The many extinctions that have been triggered or aided by climate change and the movement of tectonic plates, while individually tragic, are an unavoidable part of the story of evolution. The other side of the coin is that the destructive power of these events has opened up many opportunities for evolutionary change.

Ordinary and extraordinary earthquakes and eruptions help to move the continents along, but for the most part they have little effect on evolution. At the extreme end of this distribution, however, lie geological catastrophes that have a truly extreme impact. At least five times in the history of multicellular life, the Earth has been hit by disasters so immense that even many widely distributed species have been driven extinct. As we saw earlier, the most dramatic of these extinction events occurred 250 million years ago, not long before the breakup of Gondwanaland. Straddling the boundary between two great geological periods, the Permian and the Triassic, it is known as the Permo-Triassic extinction.

There seem to have been two events, closely spaced in time that contributed to the extinction. They involved massive volcanic eruptions so severe that the chemistry of the atmosphere and the ocean was changed. The eruptions, which racked a vast region of what is now Siberia, covered 2.5 million square kilometers with lava deposits as much as 3 kilometers thick.

Those events killed most life on the Earth. Some of the lucky survivors were given unprecedented opportunities, however. It was not long afterward that both dinosaurs and mammals appeared.

These new groups of organisms were part of a general long-term trend toward greater diversity worldwide.[11] It is striking that the numbers

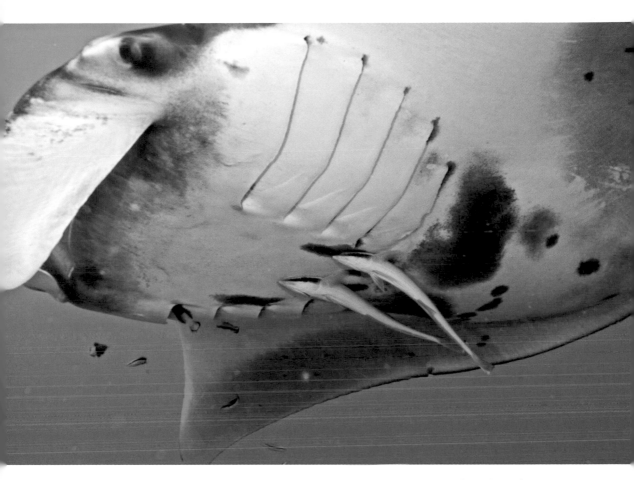

Figure 51 At the Pohnpei cleaning station tiny wrasses can be seen darting in and out of a manta's gill slits as they remove parasites. Just behind them are two sharksuckers, *Echeneis naucrates* (relatives of the more familiar remoras), who are using the suckers on the tops of their heads to hitch a ride on the ray.

of marine animal genera (groupings of species) for which we have a good and complete fossil record have gradually increased (with some temporary retreats) from a low of about 200 at the end of the Permo-Triassic event to about 6,000 today. These increases are paralleled by increases in the numbers of animal genera that live on the land. Plants have also increased in diversity, though we are less certain about their numbers because the plant fossil record is not as complete. It is clear from these trends that catastrophes small and large have not halted evolutionary diversification.

And mammals, including us, have benefited from these growing numbers of evolutionary opportunities.

Mammals arose from lineages of animals called therapsids, which were common both before and after the Permo-Triassic event. Therapsid fossils are plentiful, but until recently the fossil record of their mammalian descendants that lived at the time of the dinosaurs was sparse. But over the past twenty-five years improved paleontological methods have revealed that the mammals during the Age of Dinosaurs were almost as diverse as the dinosaurs themselves. About 310 mammalian genera have been discovered dating from that time, more than half as many as the 547 genera of dinosaurs currently known. The mammals from the Age of Dinosaurs ranged from little mouse- and shrew-like animals to much larger freshwater swimmers that looked like a cross between otters and beavers, along with fierce little wolverine-like creatures that could dine on baby dinosaurs.

Then, 65 million years ago, a meteorite or comet 10 kilometers in diameter hit the Yucatan Peninsula. It is now generally concluded that it was this impact that wiped out the dinosaurs and set the stage for the mammals.

The mammals were amply prepared for their big moment. Even though they were for the most part small and shy, they were so diverse that ten distinct major lineages of them managed to survive the meteorite impact.

The surviving mammals inherited a grim, ash-covered world. They would not have been able to repopulate that world so effectively if they had not been able to draw on the diversity that had already been produced by more than a hundred million years of earlier mammalian evolution. This diversity explains why, less than ten million years after the impact, mammals as different as whales and bats had evolved to fill ecological niches left vacant by the dinosaurs and seagoing reptiles.

Catastrophe in progress

During the past quarter of a billion years more species have appeared than have gone extinct. There are many reasons for this increase in the number of species. They include the emergence of new kinds of symbiotic relationships between different species, and increases in the numbers of predator–prey, host–parasite, and pollinator–plant interactions. We will encounter some of these processes in the next chapter, as we explore how speciation actually happens. It is this positive side of the Earth's tumultuous history, the tendency for diversity to increase even in the face of extinction, that can give us hope as we look at the future of our planet.

But we are hitting a giant speed bump on this road. We humans are causing a great wave of extinctions. This human-caused catastrophe, unlike the meteorites and vast volcanic eruptions of the past, is in a category of its own. The present time is the first time in the history of our planet that an intelligent species has appeared. We have betrayed our own intelligence by trashing the environment and sending other species into oblivion. As a species we have already wreaked far more havoc than the Toba eruption. We are fragmenting irreplaceable habitats, diverting more than half of the planet's biological resources toward the feeding and clothing of our exploding population, poisoning our environment with chemicals, and altering the planet's very climate.

But the disaster that we are causing is different from the millions of disasters that preceded it. The meteorite that wiped out the dinosaurs was unable to say "Oops, sorry!" and swerve to miss our planet. We, in contrast, should be able use our intelligence to control our numbers and behavior, in order to halt and reverse the damage that we have done. Will our grandchildren be able to look back and say proudly, "Yes, they swerved in time"?

4
Crucibles of Speciation

Figure 52 This stream, in the heart of Madagascar's Ranomafana National Park, has opened up a rare opportunity for sunlight to penetrate the dense rainforest. Many ecological niches provided by the forest have produced an explosion of speciation.

The steep trails and wet, slippery terrain of Madagascar's Ranomafana National Park are hard to negotiate when it rains, which is most of the time. I was struggling up one wet incline through a stand of giant bamboo when Roland, my guide, grabbed my arm and pointed upward. There, almost invisible in the thick tangle of bamboo leaves and stems that shielded us from the sky, were two small brown animals. They peered down from a height of about fifteen meters, directly over our heads.

The light around us was green and dim. There was no time to lose—the animals could soon vanish. I lay down flat on my back in the mud and aimed a telephoto lens straight up, much to the distress of my wife who was envisioning all the extra laundry involved. The flash, aided by a fresnel lens to concentrate the light for telephotography, gave me just enough light for a few pictures before the animals retreated behind a curtain of leaves.

Roland thought that he had spotted golden lemurs. This species, among the rarest of mammals, was discovered in 1984 by Patricia Wright of New York State University at Stony Brook. She came across them during an ultimately successful search for another lemur that had been thought to have gone extinct, the greater bamboo lemur.[1] The golden lemur, too, is close to extinction. There may be a thousand of them left, but fewer are seen every year.

When I looked at my pictures later, it was clear than we had not seen golden lemurs at all. Like the animals I had photographed, the golden lemurs, *Hapalemur aureus,* are marked by soulful dark-rimmed eyes. But golden lemurs have lighter, yellower faces, and smooth fur. The lemurs that we had glimpsed had dark faces with light raised eyebrow patches, giving them a surprised look. They also had a rich, curly brown coat quite unlike that of any other lemur. Another giveaway was a prominent light stripe running along their thighs. Instead of the golden lemur we had seen the eastern woolly lemur, *Avahi laniger.* It turned out that we were lucky even to have done that, for these lemurs are rare, usually nocturnal, and only occasionally glimpsed during the day.

In the last chapter we saw how the forces that have shaped our planet have provided plentiful opportunities for new species to evolve. But what exactly is a species? We have given names to close to one and a quarter million animal species and almost 300,000 plant species. Do all these named species have any objective existence, or are they simply artifacts of our tendency to compartmentalize the world in order to make sense of it? Is there really a difference between the golden and the woolly lemur, and if so what is it? And does the difference matter?

In this chapter we will see that species are real, and that they are an inevitable product of the interaction between the gene pools of populations and the evolutionary forces that act on them. And we will see that the total number of species that can live in a given environment depends on how complex, energy-rich, and stable that environment is.

How lemurs have become many species

In Chapter 3 we saw how India split off from Africa over 200 million years ago and began to move rapidly (at least in tectonic terms) northward toward the continent of Asia. Ninety million years ago the island of Madagascar split off from India in turn, and was left behind as India headed for an immense collision with the world's most massive tectonic plate. This vast collision would produce the Himalayas.

After Madagascar split off from India it was marooned in the middle of the Indian Ocean. It initially carried a freight of animals and plants that had originated while Gondwana was still an intact continent. But many of its first inhabitants were lost to extinction, and many of the ancestors of the present-day animals and plants of Madagascar were carried to the island more recently from Africa.

All the forty-nine lemur species today, and an unknown number of lemur species that have gone extinct, are descended from a tiny group of ancestral primates that somehow made their way to the island from Africa about 60 million years ago. They probably clung to trees that had been washed out to sea by a giant storm and then blown to a lonely Madagascar beach—rather like the animal escapees from the Central Park Zoo that made a more recent landfall on the island in the hopelessly inaccurate (but decidedly funny) animated movie *Madagascar*.

Figure 53 These eastern woolly lemurs, *Avahi laniger*, live in the deep shade of the high forest canopy and are difficult to find during the day. We mistook them for golden lemurs, until I examined my photos.

Clues that this immigration from Africa happened, though not the details of how it actually took place, can be seen from the fossil record. A few lemur-like remains from sixty million years ago have been found in East Africa. Although lemurs subsequently went extinct in Africa they were able to thrive on Madagascar, probably because there were fewer predators. Crocodiles, recently extinct on the island, must have been a constant danger. But the only mammalian predators that made it to the island along with the lemurs, aside from humans and the animals that we have introduced, are the fossas, distant relatives of the African mongooses.

Figure 54 The fossas, *Cryptoprocta ferox* and its relatives, are the only current predators of lemurs except for humans and the species that we have introduced. With its orange eyes and wicked teeth, this fossa in Andranomena Reserve in western Madagascar looks like a nightmare out of a Hieronymus Bosch painting. Fossas that were much larger and undoubtedly even more fearsome once thrived on Madagascar but are now extinct.

The first lemurs to arrive in Madagascar found themselves on an island populated by an eclectic collection of other accidental arrivals that have since evolved into species found nowhere else on Earth. On my travels around the island I encountered leaf-tailed geckos, wildly decorated Plataspid bugs, colorful chameleons, and shy hedghog-like tenrecs, along with many other wonders.

All these creatures have been dignified with distinct genus and species names. But, as we saw, the distinctions among closely related species are sometimes difficult to draw. The bamboo-eating lemurs of these dense forests are easy to misidentify, especially in the near-darkness of the forest even

Figure 55 A leaf-tailed gecko, *Uroplatus*, with an uncanny resemblance to a dead leaf, sits motionless on a dry branch in Ranomafana. The gecko's head, with its pink eyes, is on the lower right.

in the middle of the day. Indeed, Patricia Wright had initially thought that her new species of golden lemur was simply a color variant of the greater bamboo lemur that she had been searching for. Only after catching one of these new lemurs, measuring it carefully, and getting DNA samples was she sure that she had found a new species.

During our excursions into the forest we were also able to spot the supposedly extinct greater bamboo lemurs. Wright had rediscovered them on a subsequent expedition, thus rescuing them from apparent oblivion. She was able to use the publicity about her new species of golden lemur, along with the dramatic resurrection of the greater bamboo lemur, to persuade the government to declare Ranomafana a national park.

Madagascar, it seems, was an ideal place for speciation to begin among the animals that arrived there. But what drives the evolution of one species to become two or more? To find out, we must examine the ecology of some of Ranomafana's lemurs.

Figure 56 Brightly colored Plataspid bugs group together on a branch in Ranomafana Park.

Lemurs and their ecological niches

Ranomafana was established in 1991, only the fourth park to be set aside on this fragile island. The park is 435 square kilometers in extent, an area two-thirds the size of Florida's highly impacted Everglades Park. Like the Everglades, it is besieged on every side by human activity. During our visit the nights flickered with fires that had been set by farmers just outside the perimeter of the park.

The park extends from dense lowland forest to higher altitude cloud forest and is drenched by three meters of rain a year, a flood that feeds the sources of twenty-nine different rivers. It is home to twelve species of lemur, ranging from the greater bamboo lemur to the tiny brown mouse lemur and the utterly strange Aye-Aye.

The major factor responsible for the diversification of Ranomafana's lemurs into these different species is the diversity of food in the forest—a factor that is also behind many other cases of speciation. Greater, lesser, and golden bamboo lemurs are extreme dietary specialists. They eat different parts of Madagascar's endemic giant bamboos. Some of these parts, like the young shoots and the tender bases of the leaves, are tempting and nutrient-rich. Others, like the woody stems, are more daunting.

No other animals eat the bamboo, because parts of the plants are loaded with cyanide. There is enough of this poison in the plants to kill you or me, and indeed enough to do in most species of lemur.

Cyanide works by interfering with cellular respiration. We are able to use oxygen to burn our food and release energy because each of our cells contain little descendants of free-living bacteria called mitochondria. We will learn more about these supremely helpful creatures in Chapter 7, but for the moment

Figure 57 This hedgehog-like animal is a lowland yellow-streaked tenrec, *Hemicentetes semispinosus*. It lurks in the leaf litter of Perinet Forest in eastern Madagascar.

we only need to know that cyanide can instantly shut down our mitochondria, causing all the rest of our cellular metabolism to shut down as well.

The bamboos in this Madagascan forest, like many other plants around the world, make chemical compounds that contain cyanide. These compounds protect them against attacks by insects and browsing animals. The bamboos must protect their own mitochondria from cyanide poisoning, so they make their cyanide molecules harmless by attaching them to sugars. The combined molecules act like time bombs. When they are eaten by insects and animals they break down and the cyanide is released.

Some lemurs have evolved the ability to eat this deadly poison without being harmed. All three bamboo lemurs at Ranomafana can tolerate small amounts of cyanide in their diets, but the golden lemur takes the prize for being the most resistant. Only the golden lemur can eat the deadly shoots of the most cyanide-laden bamboo species. As these lemurs munch through the tender shoots they ingest enough cyanide each day to kill a similar-sized cyanide-sensitive animal a dozen times over.

Nobody knows how the golden lemur accomplishes this feat, but its droppings are rich in cyanide, showing that it somehow avoids absorbing this highly poisonous substance through its digestive tract. This ability has enabled it to carve out an ecological niche denied even to the other two bamboo-eating lemurs with which it shares the forest. The greater and lesser bamboo lemurs are forced to eat woodier, less nourishing parts of the plants because these parts have much less cyanide.[2]

The food-based ecological niches occupied by these three species of bamboo lemur are quite distinct from those of the other lemurs in the forest. The wooly lemurs that I had taken a mud bath to photograph prefer to eat the leaves and bark of broad-leafed trees, and the other lemur species have their own specialties.

Cyanide is not the only reason that the bamboo lemur species preferentially choose different tissues to eat, but it is an important factor that helps

Figure 58 (*opposite*) A greater bamboo lemur, *Prolemur simus*, noshes on its favorite food in Ranomafana's bamboo forest. Thought to be extinct, it has been rediscovered by Patricia Wright, Madagascar's "lemur lady."

to define their ecological niches. If there were only one species of bamboo in the forest, producing the same amount of cyanide in each of its tissues, instead of a mix of species and subspecies that have different amounts of cyanide in different tissues, there might not have been ecological room for three species of bamboo lemur.

The appearance of new species

How did these different lemur species first arise?

The process of speciation starts whenever a group of animals or plants is selected to live in a particular ecological niche, but this is only part of the story. Unless this group is isolated from other groups so that it cannot exchange genes freely with them, it will not become a species.

Let us suppose that a group of lemurs colonizes some new patch of forest, such that this group is completely separated physically from other lemurs. Then, over time, it will become better-adapted to its new ecological niche. There are two reasons for this. First, the genetic variation already contained in this group will be shuffled to produce new combinations of genes each generation. Some of the combinations will confer high fitness under these new environmental conditions, and the more favorable of the genetic variants will increase in frequency. Second, as a result of mutations, the gene pool of this group will begin to accumulate more and more genetic differences from other lemur groups. A few of these mutations will be favorable and increase in frequency because of selection, while other mutations that have less selective effect will replace older alleles in the population simply by chance. The result is that the group's gene pool will become more and more different from the gene pools of other closely related lemurs.

Eventually the isolated group of lemurs will become so genetically different from other lemur groups that if the physical barriers that separate them should disappear, allowing them to mix, hybrids with the other groups will be sterile or unable to survive. This is because combinations of genes that work well within the respective populations will come together for the first time in the hybrids. Some of these new gene combinations will adversely affect hybrid development or behavior, such that the hybrids may not survive or may be sterile. This isolated group of lemurs, which by this time may look and behave differently from other lemurs, has now achieved the status of a new species.

Biologists define species as groups of organisms that carry sets of genes that prevent them from exchanging genes with other groups. After its long period of isolation, our group of lemurs has diverged so far from the others that it fits the biological definition of a species.

This type of speciation is called *allopatric* or "other country," because it results from isolation. Allopatric speciation may take a long time. But, surprisingly, speciation may speed up when the population that will become the new species is not completely isolated, so that there is still some fraternization between it and other groups.

Recall the Evolution Canyon in northern Israel that we explored in the second chapter. There we saw that if selection is strong enough it can maintain genetic differences between members of a species on the two sides of the canyon, even when there is substantial gene flow between them.

Something similar probably happened in the Ranomafana rainforest. As two groups of lemurs began to specialize on different food sources, hybrids between them were poorly adapted to utilize either source. The hybrids would not have survived, or would have had few offspring of their own. Hybrids between diverging groups of bamboo lemurs in the Ranomafana forest were at an especially great disadvantage, because the available food could be potentially lethal to them. They might have been tempted to eat bits of bamboo that carried a greater load of cyanide than their bodies could withstand.

When hybrids are unfit there will be strong selection for genetic *isolating mechanisms* that tend to prevent members of the two groups from hybridizing in the first place. For example, there may be selection for different courting and mating behaviors. Because of these different behaviors, members of one group will tend to ignore those of another. The lemurs that do not seek mates across the boundary of their group will be at a great advantage, because none of their offspring will be unfit hybrids.

As alleles for these isolating mechanisms are selected for and accumulate in the groups' gene pools, the groups will no longer be able to exchange genes, even though in theory they could. And, like our physically isolated allopatric group of lemurs that gradually accumulated genetic differences until they were unable to exchange genes with other groups, these groups will also now fit the biological definition of species. But unlike allopatric groups these potentially interbreeding groups have been driven apart by strong natural selection for isolating mechanisms.

Different variants of this kind of speciation process have been given the names *sympatric* or *parapatric* ("same" or "adjacent" country). This kind of speciation can happen relatively quickly, because the alleles that confer isolating mechanisms are subject to strong natural selection.

Much of the Ranomafana forest has been destroyed by farmers and loggers. But before that recent wave of destruction the forest was extensive enough to hold a wide variety of ecological niches, and each of those niches in turn was large enough to support good-sized lemur populations. It was that earlier forest, not the tiny remnant that exists today, that made the speciation of the lemurs possible. When it comes to speciation, the size and complexity of the ecosystem plays an essential role. We will explore in the next chapter how the vast rainforest ecosystems of the tropical Americas and Southeast Asia have been among the greatest generators of species diversity.

The adaptation of one species to an ecological niche can open up more niches, which can be occupied by new species in turn. For example, nothing is greeted with more enthusiasm by the roistering insects of a rainforest floor than the arrival of a fresh animal turd. But golden lemur excreta, loaded with cyanide, are off-limits to most insects. The droppings of the golden lemur are eaten exclusively by a dung beetle that is also able to survive high levels of cyanide. Their turds have become the private property of this highly specialized beetle.[3]

These dung beetles have also been selected for isolating mechanisms, so that they do not mate with beetles of other closely allied species. Hybrid offspring would be unable to eat the cyanide-rich droppings, and they would not be able to utilize the droppings of other animals because other specialized beetle species would drive them off.

Such accumulations of nested ecological niches are characteristic of the world's richest ecosystems. They explain these ecosystems' amazing diversity, and also contribute to their fragility. If conditions change, so that a species in such an ecosystem loses its niche, its demise will drag other specialists down with it.

Some of these species will probably remain unknown to science, and will perish before we even have a chance to meet them. The bamboo lemurs of the Ranomafana Forest have a unique large intestine, its inner surface studded with bulges. Bacteria inhabit these bulges, but we know little about them, just as we know little about the crowds of bacteria that inhabit the crops of hoatzins. Do these bacteria help the lemurs extract the nourishment from the woody pulp of the bamboos without exposing them to the dangers of cyanide poisoning?

We may never know. The flickering fires that lick away at the outskirts of the fragile Ranomafana forest may soon blaze through the park and sweep away the lemurs and the many other species that depend on them. Then this magical world, built by speciation upon speciation, will be no more.[4]

Why some ecosystems are so rich in energy, ecological niches, and species

All of our planet's living systems ultimately depend on the sun. It is true that some unusual deep-sea ecosystems are powered by energy-rich molecules such as hydrogen sulfide that are released by the Earth's volcanic activity and that are used by specialized microbes, but most of the creatures that inhabit these black oceanic depths are descended from organisms that once basked in the sun far above.

Why are some ecosystems so rich, and how can they generate and support so many species? To understand this, we must look at the flow of energy in different ecosystems.

Although the sun is the ultimate source of energy, it is utilized in different ways in different ecosystems. Much of the intense sunlight that bathes the clear waters of tropical oceans is wasted. The few tiny animals and plants that float there are limited, not by energy, but by a shortage of mineral nutrients. The water is beautiful, blue, and crystal clear, but it is in effect a desert. Nitrogen, calcium, phosphorus, and iron, all essential to life, are in short supply.

Figure 59 The underwater world abounds with nested ecological niches. A pink anemone fish, *Amphiprion perideraion*, retreats into the protection of the anemone *Heteractis magnifica* on this reef off the east coast of Malaysia. Were the anemones to disappear, the anemonefish would be endangered unless they could swiftly adapt to a world that no longer offered such protective havens.

Coral reefs do much better than floating organisms, even in such desert waters. They thrive because they are built on a solid foundation of older reefs. The coral polyps, aided by the algae that live inside them, gain energy from the sun. They can also extract essential minerals from the older dead polyp skeletons on which they rest, and from the drifting and swimming plankton that they catch. The anchored polyps do well because they do not need to expend much energy to catch these swimmers and drifters. They can simply sit and wait, rather than being forced to swim around in the water and frantically chase their prey.

The solid, highly structured bed of the coral reef provides endless opportunities for additional organisms to find and occupy new ecological niches. All these niches are ultimately based on the corals themselves.

On a typical reef, bumphead parrotfish crunch their way through corals using strong lattice-like teeth. They then grind and digest the coral fragments in their muscular stomachs. Other fish swim close to see what the bumpheads will scare up. Colorful shrimp swarm over the reef, catching small creatures that hide among the corals. Wentletrap snails attack the corals directly, sucking the life out of them.

The corals in open-ocean reefs are limited by shortages of essential nutrients, but even so they support a riot of life that occupies many nested and interdependent ecological niches. But some tropical communities have extra resources that open up even more ecological niches and make them even more likely to play a role in the evolution of new species. Recently I approached the source of the unusual richness of one of these tropical communities through a series of stages.

The northeast part of the Indian Ocean, bounded by the Andaman Island chain on the west and the coasts of Thailand and Myanmar on the east, is called

Figure 60   This reef, off the eastern end of the Spice Island of Misool, just south of New Guinea's westernmost tip, is one of the lushest I have seen. Covered with soft corals, it is ruled by a lordly yellowbanded sweetlips (*Plectorhinchus lineatus*). Open ocean reefs that are far from land are not usually this exuberant, because the water is lacking in essential nutrients, but the waters around Indonesia's Spice Islands carry a rich supply of both inorganic nutrients and plankton.

Figure 61 The activities of bumphead parrotfish provide many opportunities for nested ecological niches. This bumphead, *Bolbometopon muricatum*, is cruising the Blue Corner, an underwater point that projects out from the reef surrounding the island of Palau in Micronesia. Above its head a cornetfish is poised like an arrow. The cornetfish darts forward and catches the creatures that are disturbed as the bumphead eats it way through the corals.

the Andaman Sea. Warm, sheltered, and relatively shallow, the Andaman Sea swarms with shrimp—and also swarms with Thai shrimp boats. At night the lights of their lanterns form a line that spans and defines the dark horizon.

Despite this overfishing, life beneath the surface of the sea is enormously varied and abundant. The south Andaman Sea waters off the Thai coast are a favorite place for scuba divers, because they are clear and have a fine collection of sharks and rays. As I was preparing for my first dive at Ko Bon in Thai territorial waters, the cry went up: "Whale shark!" We stuffed snorkels into our mouths and leaped into the water. I was a little behind the others, because I had to grab my camera.

Figure 62 Bumphead parrotfish tend to be shy, so I was lucky to grab this close-up on a reef off peninsular Malaysia's eastern coast. The remarkable fused teeth of this largest of all parrotfish, embossed in a crisscross pattern like a coarse file, can crunch easily through the most massive hard corals.

When I entered the clear blue water I found an immense whale shark swimming straight towards me. These gentle wide-mouthed giants are harmless plankton feeders, but their size makes them awe-inspiring. They are the largest living fish, and can be as much as fifteen meters long. This one veered slightly as it approached, and slid past me only a meter away. A large calm eye gave me the once-over. With a mighty sweep of its tail the shark disappeared into the blue.

Sharks, rays, and moderately diverse reefs are the norm in Thailand, but as you sail north into Myanmar waters the mix of organisms changes.

Getting into this area is not easy. We anchored off the undistinguished town of Kawthaung on the Thai–Myanmar border to await a Myanmar customs official. He was dressed in a uniform straight out of the old Marlene Dietrich movie *Shanghai Express*. As dusk fell we entered into complex negotiations involving passports and large sums of money. Satisfied at last, he swung down into his launch and departed.

We sailed northwest that night, and the next morning found ourselves well inside the region of the Andaman Sea claimed by Myanmar. First light revealed a set of jagged reddish rocks that jutted out of the sea. There were three of them, and divers had with great inspiration named the site the Three Islets.

These steep rocks formed part of the scattered Mergui Archipelago, a mostly uninhabited set of hundreds of tropical islands strewn up Myanmar's long southern coast. Some of the Mergui islands are quite large and heavily forested, and some even support a few wild elephants.

Above water the Three Islets were bleak and vegetation-free. The only signs of life were a few fast-moving ghost crabs that fled into rock crevices at our approach. But underwater the islands' slopes were a riot of color.

Figure 63 Corals are eaten in a wide variety of ways. Here wentletrap snails, *Epitonium billeanum*, feast on orange cup coral polyps in a reef off Thailand. As they feed the snails are laying a mass of yellow eggs, most of which will be eaten by other predators.

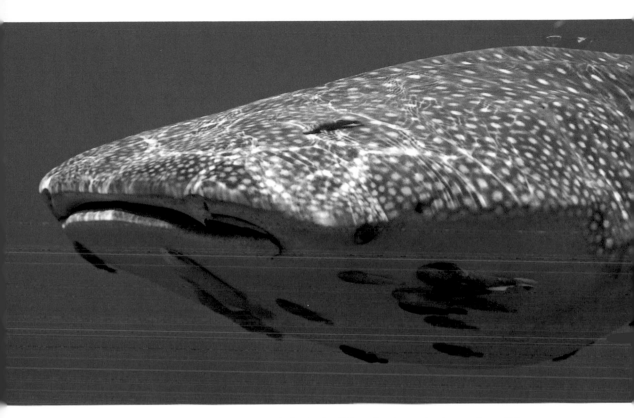

Figure 64 I encountered this huge whale shark, *Rhincodon typus*, in clear waters off the coast of Thailand. Clusters of remoras, *Remora albescens*, are attached to its chin by the suckers on the tops of their heads. This shark is a plankton eater, and is moving north into the ever richer waters of the Andaman Sea.

The visibility was poorer here than in the blue waters further south in Thailand. But the worsening visibility was more than compensated for by the sheer diversity on display. Every square inch of the reef was covered by orange and green cup corals, sponges, soft corals, and sea fans. The soft coral growth was so plentiful that predators of the corals such as the wentletrap snails were more numerous than I had ever seen them. I swam forward and touched one of the rare patches of bare rock on the reef, thinking as I did so that the hermetic land of Myanmar was the only one of the world's countries that I had touched for the first time while underwater.

Camouflaged scorpionfish waited patiently for incautious passersby on many of the reef's ridges, the only giveaway to their presence being their

Figure 65 Three *Chromodoris annulata* nudibranchs mate on a crowded part of Myanmar's Black Rock reef.

dark unblinking eyes. Even the cuttlefish that hovered over the reef and the octopuses lurking in hollows were intricately patterned and colored, covered with spotted retractable bumps that imitated the blazing palette of colors on the reef itself. And everywhere in this riot of life were nudibranchs, the little molluscan jewels of the sea.

Each anemone on the reef had a full complement of anemone fish and glittering porcelain crabs. The real estate was so crowded that many crevices were inhabited by two or even three moray eels. Food sources were so plentiful that often two species of moray shared the same hole, even though on most reefs morays will defend their holes fiercely.

The difference between this crowded world of the Mergui reefs and the simpler reefs that we had left behind in the Thai waters to our south was dramatic, and it increased as we moved further towards the coast of Myanmar. Visibility went down as we sailed to the north, but the sheer roistering exuberance of life grew even greater.

Away from the solid reefs, however, the story was different. Small fish were abundant, but the large schools of snappers, bannerfish, and triggerfish

Figure 66 This Durban dancing shrimp *Rhynchocynetes durbanensis* rules a tiny part of the Three Islets reef.

that we had seen in the Thai waters were absent in Myanmar because of over-fishing. The great diversity of the Mergui reefs should have supported many more fish than the Thai reefs, but the fish populations had been decimated.

The Thai reefs on which so much of the country's tourism depends are informally policed by the dive boat operators. Fishermen venture into the waters around Thailand's Similan Islands at their peril. But there is no such protection in Myanmar.

The loss of fish that are at or near the top of the food chain will eventually cause a cascade of damage to Myanmar's reefs. These fish, the secondary and tertiary consumers, are essential for maintaining the reef's diversity. If particular species of corals, algae, and small animals on the reefs become over-abundant, the grazing fish will find them more easily and prey on them preferentially. This allows rarer species to multiply. Then, when those formerly

Figure 67 A white-eyed and a fimbriated moray share a single hole in the crowded Mergui reef at Three Islets. Such cooperation over real estate is only seen when food is sufficiently plentiful that morays do not need to defend their home territories.

rare species become too abundant in turn, the predators switch their attention to them. Such constantly shifting patterns of predation ensure that many species of smaller organisms can coexist simultaneously. But when the top predators are removed their fastest-growing prey become overabundant, and the reef suffers.[5]

The source of the reef's richness

The reefs became more crowded with life as we moved north because we were gradually moving closer to the delta of the Irrawaddy River. This great river, arising in the glaciated highlands where Myanmar, India, and China come together, gathers nutrients as it flows south and pours them out through its wide flat delta.

The combination of tropical sunlight and nutrient-laden water feeds the corals of the offshore reefs. During my night dives in the Andaman Sea

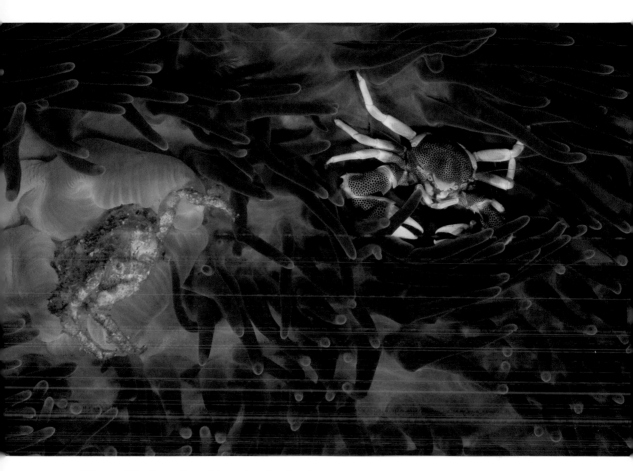

Figure 68 These Porcelain and reef crabs share an anemone.

I could see the effects clearly. The clouds of plankton drifted like snow in the beam of my lamp, their abundance showing up far more clearly than during the day. Great beds of orange cup corals opened up like flowers, waving their tentacles to gather in the small animals and plants as they swam by.

A few days later I traveled up the Irrawaddy from Myanmar's capital, Yangon, to the charming village of Tawnte. This was fifteen months before Cyclone Nargis hit the region in May 2008, killing 150,000 people. On this bright late January day in 2007 such a disaster seemed unimaginable. Our boat sailed up a beautiful river, far cleaner than the Ganges to the west or the Mekong to the

Figure 69 Yellow sweepers, *Parapriacanthus ransonneti*, swoop through a cavern beneath one of the Mergui islands. Such small fish survive in abundance on the reef, but populations of larger fish are rare.

east. Rice fields came right down to the garbage-free banks, and each neat village that we passed was dominated by a gleaming golden stupa.

Because there has been so little industrialization along the Irrawaddy, the water flowing into the Bay of Bengal and the Andaman Sea is not yet loaded with the toxic chemicals that fill the other rivers of the region. But there is one aspect of the Irrawaddy ecosystem that has been strongly affected by human activity. Fishermen now toil on its banks with very little result. Further up the river, near the temple-filled valley of Bagan, I watched teams of fishermen

146

spread their heavy nets in the river's shallows. They waited patiently, then hauled in the nets to yield a sparse handful of wriggling small sardine-like fish. The Irrawaddy once teemed with fish, but dam-building and overfishing have emptied the river.

The industrialization of Myanmar has been suppressed by its current government, a cruel and ignorant collection of generals and their greedy relatives. Ironically, this may be the temporary salvation of the teeming reefs of the Mergui Archipelago. By the time the agricultural runoff reaches the Mergui islands, hundreds of miles from the Irrawady's multiple mouths, it is capable of nourishing a wide variety of plankton. More intensive agriculture and industry will inevitably result in a witches' brew of nitrogen-rich runoff mixed with industrial waste that will certainly disturb this balance in the future.

A voyage through the Mergui islands today reveals a wonderfully rich but fragile ecosystem. Invisible to the casual visitor, it is immediately apparent to anyone who ventures beneath the ocean's surface. Because of the great clouds of plankton nourished by the river, there is plenty of room for many species of animals and plants to carve out specialist ways of feeding and establishing territories.

But a rich flow of plankton is not enough to explain such diversity. Ecosystems in the Arctic and Antarctic are rich in plankton but support far fewer species than the Mergui Archipelago. To support diversity throughout the food chain the plankton themselves must be diverse, providing a wide variety of potential food sources and their associated nested ecological niches. In Mergui this diversity is supported by ample solar energy and inorganic nutrients. In the next chapter we will see how, even at the level of the single-celled plankton on which the rest of the food chain depends, ecological and evolutionary forces have combined to produce and maintain a diversity of species.

Ecological niches

A principle of ecology that was first proposed early in the twentieth century states that two species cannot occupy the same ecological niche at the same time. This *niche-exclusion principle* has been supported by simple laboratory experiments, but in the real world of a coral reef or a rainforest things become much more complicated. In such a complex world, what exactly is an ecological niche? Where does it begin and end? How much do two niches have to differ in order to support different species?

Peter and Rosemary Grant and their colleagues have examined the nature of ecological niches in the real world, and their relationship to speciation, in the Darwin's finches of the Galàpagos Islands.[6] The islands of this mid-ocean archipelago range in size from the ecologically diverse Isabela, the largest island at 4,640 square kilometers, down to tiny rocks that barely stick out from the sea. The Grants found that an island's size determines the number of species of finch that the island can support.

On the larger islands of the archipelago there tend to be large numbers of a wide variety of seeds available, so that each species of seed-eating finch

Figure 70   Fisherman deploying their nets on the Irrawaddy River near Bagan in Myanmar.

Figure 71 The unimpressive result.

has the luxury of being able to specialize on one particular size of seed. As a result, several seed-eating species can survive and thrive. But smaller islands are so small that the amount of food they produce is only enough to support one finch species. If two or more species of finch lived on these little islands, each species would have such a small population size that it would be threatened with extinction when times became hard. Thus, before long, only one species would survive.

The plants on the small islands produce a range of seed sizes, just as they do on the large islands, but there are simply not enough seeds of each size class to maintain a specialist finch population. The finch species that can survive on the smaller islands are those that are capable of eating a wide variety of seeds.

Members of the seed-eating finch species *Geospiza fortis* that live on a big island such as Isabela are able to specialize on the island's abundant large seeds. They have evolved large powerful beaks. But members of the same species living on the tiny island of Daphne Major have a middle-sized beak, suitable for cracking a wider range of different seeds. Daphne Major is only one-third of a square kilometer in area, and it is this generalist version of *G. fortis* that makes up most of the finches on the island.

The species *Geospiza fuliginosa* on large islands has specialized on small seeds, and has a small beak as a result. On the little cluster of three islands off the coast of Isabela known as Los Hermanos, each of which is about the size of Daphne Major, *G. fuliginosa* is the predominant finch species. But on at least one of these little islands *G. fuliginosa* has evolved a more middle-sized, generalist beak, like *G. fortis* on Daphne Major.

Differences in foraging behavior have evolved as well, so that the birds on a large island behave differently from birds of the same species on a small island. The result is that big islands, with their abundant resources, can support more species than smaller ones. On the larger islands it is possible for species to be narrow specialists. On the smaller islands, where such narrow specialists would be unable to obtain enough food, the species that are the most resourceful and best able to adapt to a broad rather than a narrow ecological niche are the ones that have won out.

Peter and Rosemary Grant were able to examine the individual factors that have shaped the evolution of the relatively small number of finch species in the Galápagos Islands, because the ecosystem of these islands is relatively simple. The Mergui Archipelago and other tropical ecosystems that teem with life are evolutionary hothouses, battlegrounds where natural selection interacts with numerous ecological niches to spew forth a stream of new species. Such ecosystems are so rich in available energy that even highly specialized ecological niches provide sufficient resources to maintain a species.

I realized, as I swam among the crowds of fiercely competing species on Myanmar's reefs, that I was looking at a kind of delicate evolutionary baroqueness. Perhaps, I thought, these swarms of species are akin to those decorative elaborations of furniture and décor that became more florid with each successive Capetian king during France's *Ancien Régime*.

Conditions changed in France with the fall of the Bastille and the rise of the grim and decidedly undecorative Committee of Public Safety. Talleyrand, who survived the French Revolution and many subsequent upheavals, famously said: "*Ceux qui n'ont pas connu l'Ancien Régime ne pourront jamais savoir ce qu'était la douceur de vivre.*" Those who never knew the *Ancien Régime* will never know how sweet life can be.

And now conditions are changing on the whole planet, as we play the role of Robespierre and destroy entire ecosystems. It is a truism that those who have never experienced the sheer exuberance of rich ecosystems will never realize the sweetness of life as it used to be.

Will most of these wonderful species, these evolutionary ornaments, disappear? If they do, what are the conditions under which similar species could evolve in the future? How long might it take before the planet can return to the delightful excesses of evolutionary ornamentation that still survive in the Mergui Archipelago?

In the next chapter we will explore these questions. We will see how ecological diversity is produced by natural selection, how it is maintained over time, and how it is destroyed. And we will confront a remarkable paradox. The process of evolution has endowed the living world with both fragility and robustness. Life on Earth will survive no matter what we do. Our great challenge is to see whether we can preserve enough diversity on our planet to make the lives of our descendants worth living.

5

Rainforests, Diseases, and Speciation

Figure 72 Dawn breaks over the mist-filled forests of the Danum Valley in central Sabah, Borneo. Tropical forests are some of the world's most complex ecosystems, and yet at first sight it is not obvious how they can harbor the many different ecological niches that have led to an explosion of animal and plant speciation.

Charles Darwin first encountered a tropical rainforest in Brazil's Bahia state during one of the *Beagle*'s first landfalls. His excitement spilled over into the journal he was writing:

The day has passed delightfully. Delight itself, however, is a weak term to express the feelings of a naturalist who, for the first time, has wandered by himself in a Brazilian forest. The elegance of the grasses, the novelty of the parasitical plants, the beauty of the flowers, the glossy green of the foliage, but above all the general luxuriance of the vegetation, filled me with admiration.

When Darwin visited Brazil, in early 1832, its coastal rainforest stretched almost unbroken for 3,000 kilometers. He could simply stroll to the outskirts of Bahia's largest city, Salvador, and find himself in primeval forest. But most of Brazil's coastal forest has now been destroyed. To see a comparably undamaged rainforest today one must go elsewhere.

Rainforest diversity

In 1994 I ventured for the first time into the rainforests of Peru's upper Amazon basin. I had no idea that this trip would be the genesis of a research project that would consume me for the next fifteen years.

The project began by accident. I was in Peru because I was writing a book about how human diseases have evolved over time. I had the chance to talk to many people who had experienced Peru's severe cholera epidemic, including doctors and nurses in some of the rainforest towns and settlements that had been especially hard-hit.

During the trip I seized the opportunity to visit Manù, the country's largest national park, on the theory that every biologist should, like Darwin, have the chance to wander through an untouched rainforest. The remote Manù region includes both cloud forests on the slopes of the eastern Andes and densely forested flatlands that form part of the great Amazon basin. The lowland forest is filigreed with serpentine brown rivers and dotted with oxbow lakes that the rivers have left behind as they writhe with imperceptible slowness across the landscape.

Figure 73 Thunderheads cast giant shadows across the upper Amazon at sunset. Even 3,000 kilometers from its mouth, this mighty river can easily accommodate ocean-going freighters.

Like Alfred Russel Wallace, more than a century earlier, I was astonished by the sheer diversity of Manù's rainforest. Wallace had remarked on the sheer numbers of different tree species crammed into the Brazilian rainforest, so many that one almost never finds two members of the same species growing together.[1]

Exhaustive censuses have now shown that rainforests in Borneo, Peru, and Venezuela can have almost a thousand different tree species in a single hectare—that is, in a patch of ground that measures a hundred meters by a hundred meters. The trees carry on their branches, trunks, and roots a dizzyingly diverse freight of epiphytic plants, insects, amphibians, reptiles, birds, and mammals, along with a little-explored zoo of mostly invisible creatures like bacteria and fungi.

Tropical ecologists who came after Wallace were puzzled by this species diversity. How can so many different species of tree live packed so closely together in what seems at first blush to be an ordinary, rather uniform, wet tropical environment? It is immediately obvious that there are tall trees, medium-sized trees, and short trees in a rainforest, each adapted to different amounts of sunlight. But differences in sunlight account for only a small number of ecological niches. How can a rainforest possibly encompass hundreds or thousands of different ecological niches, when it is clear that all the plants that form the base of the food chain have access to at least the necessities of life?

Although I had no background in tropical forest ecology I began to wonder whether I could apply principles of population genetics to the problem. And it was at that point that I embarked on the research project that eventually provided some answers to these questions.

The origin of the rainforests

To understand rainforests, we must begin with their history. A mature rainforest of the present day consists of an understory of small and young plants, a dense midstory of medium-sized trees, and a towering canopy of a few giant trees that have managed to break through into the open sunlight high above.

Rainforests of the past had a similar overall structure but a different mix of plants. At successive times they were dominated by seed ferns, cycads, and evergreens. One of the great events in evolution has been the recent transformation of forests throughout much of the world as they were taken over by flowering plants.

Trees and other plants that reproduce using flowers have a more sophisticated reproductive system than the cycads and evergreens that dominated forests in the distant past. Flowering plants can attract a wide variety of pollinators, which help to ensure fertilization of their eggs. But perhaps more importantly, the flowering plants use a trick called double fertilization to provide stored food only to the eggs that are fertilized. They do not waste

Figure 74 A wide variety of tree species are reflected in the seasonally flooded Marañòn River of the upper Amazon.

energy building up provisions for unfertilized eggs, as sometimes happens with other seed-bearing plants such as the seed ferns.

Although flowering plants first appeared at the beginning of the Age of Dinosaurs, 140 million years ago, they did not dominate tropical forests completely until near the end of the dinosaurs' reign. And tropical rainforests that resemble those of today seem to have appeared only after the beginning of the Age of Mammals, 65 million years ago.

Kirk Johnson and Beth Ellis of Colorado's Denver Museum of Nature and Science found traces of an early post-dinosaur fossil forest when it was exposed during excavations for an interstate highway at Castle Rock, just south of Denver.[2] The forest was thriving a mere 1.4 million years after the mass extinction that wiped out the dinosaurs. It consisted almost entirely of diverse flowering trees adapted to heavy tropical rainfall. Very few of the species of tree that lived in the Colorado forest can be traced to before the time of the mass extinction, so this forest really was something new on the planet.

In the 1980s Bruce Tiffney, then at Yale's Peabody Museum, made an astounding discovery. As flowering plants took over the forests they underwent an enormous evolutionary change. Their average seed size, which was tiny when the dinosaurs were dominant, suddenly increased by ten- to a hundred-fold at the beginning of the Age of Mammals. The average has stayed at these large sizes ever since.[3]

Why this rapid increase in size? Tiffney points out that in dense forests large trees must produce large seeds and fruit if they are to ensure that their seeds will be well dispersed by animals and birds. Large animals and birds are attracted to such large seeds and fruit, which provide the substantial packets of energy they need. The large animals will carry the seeds they don't eat to places far from the parent tree, where they have a chance to grow up into the canopy.

But, Tiffney suggests, this evolution toward larger seeds and fruit only became possible when large mammals and plentiful fruit-eating birds had replaced the dinosaurs. The dinosaurs tended to be rather unimaginative when it came to cuisine, and had less demanding metabolisms than mammals. Herbivorous dinosaurs simply chewed up entire plants, perhaps fermenting them like the animals we met in Chapter 2. They did not take advantage of the concentrated sugars, fats, and proteins that were offered by the fruits and seeds of flowering plants, in part because these offerings were so skimpy. A giant brontosaurus could not survive by nibbling on teeny-tiny fruits and seeds.

In contrast, herbivorous mammals evolved to take advantage of the goodies provided by the flowering plants, right from their origin during the Mesozoic. The small mammals, unlike the dinosaurs, could move into the ecological niches provided by these tiny but energy-rich nibbles. Seed-eating was a boutique niche, allowing only small populations of small-sized mammals

Figure 75 (*opposite*) This *Banksia prionotes* from western Australia is pollinated by a wide variety of birds, but the pollination success rate is low. The orange inflorescence matures into an infructescence, such as the one you see at the lower right. Only a few of the flowers of the inflorescence were pollinated, and each of these double fertilizations has triggered the development of large brown seed-containing follicles. The mechanism of double fertilization enables the plant to make seeds only from its fertilized eggs, enabling it to save large amounts of energy.

to coexist with the dinosaurs. If the mammals became too numerous or too large, carnivorous dinosaurs would notice them and eat them up. It therefore behooved the mammals that shared the world with the dinosaurs to keep a low profile. And this meant that there was no selective pressure acting on the flowering plants to produce bigger seeds and fruit.

Once the dinosaurs vanished, however, mammals could grow larger. Large carnivorous and herbivorous mammals were able to take the place of the large carnivorous and herbivorous dinosaurs. The herbivorous mammals were used to eating fruits and seeds, and now they needed large and rich sources of food to fuel their large bodies and brisk metabolisms. The flowering plants rapidly co-evolved with them to provide these sources of energy.

The diversities of mammals, of birds, and of the flowering plants themselves exploded at the beginning of the Cenozoic. The dinosaurs had kept a lid on all this evolutionary activity. Their demise is regretted by some (especially by small dinosaur-smitten children), but personally I would rather share the planet with intelligent, behaviorally diverse mammals, brilliantly colored flowers and birds, delicious fruit, and forests that are far more diverse than anything that existed during the Mesozoic.

Theories about rainforest diversity

In the green teeming world of the rainforest millions of organisms are living and dying, all of them subject to the unrelenting pressures of natural selection. How can we make sense of all these individual stories of life and death? How can each of these stories help to explain the diversity of the entire rainforest?

Many theories that try to make sense of the biological cacophony of complex ecosystems have been proposed. At the time I visited Manù I knew something about one of them.

In the 1970s, Daniel Janzen and Joseph Connell had suggested that high levels of tree diversity in tropical forests could be explained by the threats that surround and besiege each tree.[4] Trees, they pointed out, live in a kind of precarious equilibrium with entire collections of herbivorous animals, insects, fungi, and bacteria that prey on them.

Figure 76 A Curculonid weevil from the Mulu forest of Sarawak. These beetles use their sensitive antennae to home in on chemical signals from other members of the same species. They make holes in saplings to lay their eggs. The smaller the sapling, the more likely it is to be badly damaged by the burrowing larvae.

Janzen and Connell proposed that this swarming zoo of pathogens has its greatest effect when the tree is a tiny sapling. Most saplings die as a result of their pathogens, but if a little sapling is lucky enough to survive their depredations it will become bigger and hardier and more resistant to subsequent attacks. As it reaches adulthood it is able to withstand the great crowds of pathogenic bacteria, fungi, and insects that, along with the animal browsers and seed-eaters, tend to accumulate around it.

When the tree reaches reproductive age, it begins to scatter seeds. If the seeds fall nearby, they land in the midst of a dangerous world, filled with pathogens waiting for them. If they fall further away, out of the range of this nasty collection of predators and parasites, they have a higher chance of surviving.

Figure 77 These caterpillars, probably of Nymphalid butterflies, swarm over a leaf in the Lambir forest of Sarawak, Borneo.

Janzen and Connell visualized a process in which the seedlings of a tree are at a disadvantage when they are close to the parental tree, but at an advantage when they are further away. The result will be a dynamic, constantly shifting pattern of trees of many different species within the forest.

As a cluster of trees of a given species grows up, it accumulates the pathogens that are adapted to prey on this species. Any seeds that happen to fall to the ground within the cluster of adults are unlikely to survive because of all the pathogens and predators in the area. But if the adults can manage to broadcast some of their seeds outside the clusters, a few of these widely dispersed seeds will eventually establish new clusters. This is because they will fall in relatively pathogen-free zones. Then, as these new clusters begin to mature, they attract herbivores and pathogens and begin to run into the same reproductive problem as the original cluster.

Meanwhile, the original cluster of trees grows old and dies. Its pathogens also die, or migrate to new clusters. And now the area where the trees and their pathogens used to live is freed up to host a new cluster.

Mathematical ecologists who followed up on Janzen and Connell's original idea pointed out that such a dynamic system could maintain many species of tree simultaneously, with each tree of a species at an advantage when it is rare in a pathogen-free part of the forest, and at a disadvantage when it is part of a cluster that attracts pathogens and herbivores.

This dynamic process resembles a phenomenon called frequency-dependent selection, which has often been found to be acting on genes within populations. An allelic form of a gene may be at an advantage when it is rare, but at a disadvantage when it is common. Such alleles will spread in populations, until they reach an equilibrium point at which the advantage of rarity is counterbalanced by the disadvantage of commonness.

Figure 78 Grasshoppers are everywhere in the forests and do much damage to leaves and stems. I spotted this striking Acridid grasshopper in Gunung Gading National Park, near Kuching in Sarawak.

Just as this process of frequency-dependence can pack a population full of different alleles of a gene, the Janzen–Connell process can pack an ecosystem full of different species. Each species is at an advantage when it is rare and at a disadvantage when it is common.

The Janzen–Connell hypothesis has been tested directly in several ways, using both tropical and temperate tree species. Perhaps the most elegant experiment was carried out by Alyssa Packer and Keith Clay of Indiana University. They showed that temperate-zone black cherry trees from an Indiana forest showed exactly the kind of growth pattern that had been predicted by Janzen and Connell.[5]

Packer and Clay found that cherry seedlings thrived when they were planted in soil that had been dug from a part of the forest that had no adult cherry trees. But when they raised the seedlings in soil that was taken from areas close to adult trees, the seedlings tended to get fungal infections and die.

They then sterilized samples of soil taken from sites close to an adult tree and from a part of the forest without cherry trees. They found that the seedlings thrived in these samples of heat-treated soil, regardless of where the soil came from. This experiment showed that it was organisms living in the soil near the adult trees that were inhibiting growth. To help close the circle of proof, they used a classic approach that had been pioneered by Robert Koch, the nineteenth-century microbiologist who discovered the anthrax bacillus. They cultivated a root rot fungus from diseased seedlings and showed that it was able to kill a new set of seedlings. They had not proved that this particular fungus was entirely responsible for the deaths of seedlings in their greenhouse experiment, but they did show that it was likely to have been a contributor to their deaths.

Some tree species in tropical forests have been tested for the Janzen–Connell effect in similar experiments, though not as thoroughly or elegantly. Some of these experiments gave positive results and others showed no Janzen–Connell effect, a point that I will return to in a moment.

Figure 79 (*opposite*) Trees festooned with lianas dominate this rainforest in peninsular Malaysia. The balance between life and death that maintains diversity in such a rainforest may be invisible to the casual visitor, but it is essential for the forest's health.

Everywhere you look in a rainforest you see interactions between plants and the insects that prey on them. And everywhere the leaves of trees in the forest are stippled with the signs of fungal, bacterial, and viral infections.

Predation and infection are daily hazards of life for the plants and animals living in these complex ecosystems. Visiting photographers, however, are more likely to be affected by the defenses that plants have thrown up against these predators than by the predators and pathogens themselves. In Central America I have been bitten viciously by ants that were defending young Cecropia trees. The trees provide special chambers in which the ants can live, and benefit from their fierce aggression against any invaders. And in Sarawak's Mulu River forest I accidentally brushed against a plant that was probably a member of the Family Anacardiaceae, which includes the sumacs and poison ivy. The resulting itchy red welt on my arm lasted for months before it finally faded away. I had become collateral damage in the ongoing warfare between plants and their enemies.

Pathogens and predators can play their diversity-maintaining role unimpeded in relatively untouched rainforests like Manù. But their role is changing rapidly in forests that have been damaged by human activity.

In Malaysia's province of Sarawak in northern Borneo much of the rainforest near the coast has been clearcut to make way for oil-palm plantations. But there are still extensive forests covering the Lambir Hills, and substantial areas of forest along the Mulu River. The trees in these forests range from small bushes, caught in a time warp and hardly able to grow on the forest floor because there is so little sunlight, to mighty dipterocarps that tower eighty meters into the full sunlight. These emergent trees can produce an explosion of flowers and seeds.

Arboreal animals such as monkeys and giant hornbill birds play essential roles as eaters of fruit and scatterers of seeds. Browsing animals like tapirs and wild pigs munch on leaves and trample seedlings. Only the strongest plants survive this herbivory. But the animals and large birds that used to fill Sarawak's forests are now unrelentingly hunted by poachers from peninsular Malaysia. And the growing populations of local tribespeople now have smaller and smaller areas of forest in which to practice their traditional forms of hunting, putting even more pressure on the birds and mammals that remain.

There are a few exceptions. The native people living around Bako Park near Sarawak's capital of Kuching are now Muslim, and consider forest animals unclean. As a result the animal and bird life in and around Bako is still abundant.

The Janzen–Connell effect is unlikely to be the only factor maintaining ecosystem diversity. Recall that some tree species in tropical forests seem to be uninfluenced by this effect. There is another major theory about rainforest diversity that may account for these species. It has been given the rather clumsy name of the niche-complementarity theory, which we will shorten to N-C.[6] Like the Janzen–Connell effect the N-C theory is also based on frequency-dependence.

The N-C theory assumes that the environment is divided up into a variety of niches defined by physical components such as sunshine, water, and type of soil, and biological components such as different kinds of prey. Trees, of course, do not prey on other organisms, so a tree species is limited by the availability of different components of its physical environment. If a tree species is rare in a particular part of the forest, it is at an advantage because it basks in a plenitude of the environmental component that it depends on the most. Trees of a species limited by a particular mineral in the soil, for example, will thrive if they are not competing with other members of the same species for that resource. But as trees of this species increase in numbers they must compete more and more amongst themselves for the limited amounts of the mineral available in that part of the forest.

Tree species that are rare in the entire forest are more severely limited by their environmental requirements than species that are common, but both rare and common species are subject to frequency-dependent selection. Just as with the Janzen–Connell hypothesis, the N-C hypothesis predicts that if a species becomes locally more common in some part of the forest, it will have trouble producing saplings there. And, just as with Janzen–Connell, N-C predicts that clusters of trees should be continually appearing and disappearing in different parts of the forest. Finally, like the Janzen–Connell hypothesis, the N-C hypothesis predicts that many different species will be able to coexist simultaneously. These species will all oscillate back and forth in numbers, because they are selected for when they are rare and selected against when they are common.

When we examine data from a forest it is hard to decide among these and various other allied hypotheses, the most likely of which also turn out to have the property of frequency-dependence. My contribution was to find a way to detect frequency-dependent selection in the data from tropical forests, and to show that Janzen–Connell, N-C, or some combination of such frequency-dependent processes really are affecting many species simultaneously.

Barro Colorado Island and the dynamics of diversity

My work depended on an immense and ongoing project in tropical ecology. In 1979 ecologist Stephen Hubbell, now at UCLA, reported the results of a complete tree census of a 13-hectare dry tropical forest plot in Costa Rica. He found a complex pattern of scattered and clustered tree species, a pattern that cried out for more detailed analysis. But the dry forest was cleared for agriculture soon after he finished the study. Frustrated, he resolved never to study a plot that was in danger of being destroyed. Together with Robin Foster of Chicago's Field Museum, he proposed a daring idea to the Smithsonian Institution.[7]

Hubbell and Foster wanted to carry out a complete census of a piece of tropical rainforest, by determining the species and location of every tree above a certain size. Then they wanted to come back five years later and see what has happened to them all. Would they still be there? Would they have grown, and if so by how much? If they disappeared, which trees replaced them? Then they wanted to come back in another five years and do the same thing. The idea was to produce a body of data spanning decades that would be unprecedented in its scope and detail.

Hubbell and Foster got the OK for this immense project. They picked an almost completely untouched piece of tropical forest, half a square kilometer in size, on Barro Colorado Island in Panama's Gatun Lake. There was already a Smithsonian tropical research center on the island, where students from all over the world study tropical ecology. And the island, surrounded by the artificial lake that had been flooded between 1907 and 1913 to form a major part of the Panama Canal, was protected from development. They would not

Figure 80 Suspended on a walkway high above the Lambir forest plot, Sylvester Tan, a Bornean forest ecologist, surveys the forest that has been his life work. He knows every one of the almost 1,500 tree species in the Lambir plot, and many more besides.

come back for the next census to discover that their plot had been converted to soybeans or condominiums.

The first census, by far the most challenging, required that the 300,000 trees on the plot be located and identified. An army of students had to brave mosquitoes, chiggers, and tropical downpours for months. They also had to survive occasional confrontations with bands of collared peccaries that rooted and snuffled their way through the forest and threatened to attack anyone who got in their way.

In order to keep from going crazy by trying to measure millions of tiny threadlike saplings, Hubbell and Foster quickly decided to limit the censuses to trees and bushes that were one centimeter or more in "diameter at breast height." (This imprecise and politically incorrect measure has now been replaced by the less potentially embarrassing "diameter at 1.4 meters above the ground"—but it is still abbreviated dbh!)

The Barro Colorado plot has now been censused six times, with a seventh census scheduled for 2010. The result has been by far the most extensive set of data on a large ecosystem anywhere on the planet. And the census has proved to be so productive of exciting research that new censuses have been initiated throughout the world's tropics. Most recently, plots have been set up in the temperate zones of North America, Europe, and China. There are now more than thirty large plots, or collections of smaller ones, throughout the world, many of which have been censused more than once.

My own contribution to this immense project began when I returned to the USA from Manù. In collaboration with Hubbell and his colleagues I began to examine the forest data to see whether frequency-dependent effects are actually operating, and how many of the species in the forests are affected by them.[8]

Frequency-dependent models like Janzen–Connell and N-C predict that if trees of a given species are common in a certain part of the forest, they should be less successful at reproducing than in other regions where they are rare. Locally common trees should be at a disadvantage, and locally rare ones at an advantage.

The data from the Barro Colorado Island plot are not ideal for this kind of study, because information about the tiniest seedlings in the forests is not collected in the censuses. But I was able to use the locations of many trees that were not counted in one census but appeared in the next. Such a sudden appearance shows that the tree had exceeded the one-centimeter cutoff, which allowed it to be recruited into the census data. Ecologists refer to such *arriviste* trees as recruits.

I could count the numbers of such recruits in different parts of the forest, including parts where there are lots of adults of the same species and other parts where there are few or none. Are there more recruits where there are few adults? If so, then I would have found the frequency-dependence predicted by the Janzen-Connell and the N-C models.

The analysis was complicated by the fact that trees tend to shed their fruit and seeds in a decidedly non-random way. Parental trees tend to shower most of their seeds around their immediate vicinity. Fewer of their seeds land further away. Even if the seeds that land near the parental trees have difficulty surviving, because of all those pathogens and predators waiting to devour them, their sheer numbers mean that more of them will survive than in places where both the adult trees and the seeds of that species are less plentiful. This excess of seedlings may not be enough to maintain the cluster of trees over time, but it will be enough to confuse the analysis.

To circumvent this bias in the data, I asked what would happen if recruitment were to take place completely at random. I did this by building a series of "pretend" forests in the computer, and then comparing results from them with the way recruitment happens in the real forest.

In these pretend forests I put all the same trees into the same positions as in the real forest, but then I went to each species in turn and shuffled the recruits around at random. In the pretend forests, unlike the real forest, the chance that a given tree would be a recruit was always the average chance for that species, regardless of whether that tree was surrounded by lots of other trees of the same species or by trees of different species.

When I did this, I found that the pretend forests were clearly different from the real forest. For almost all of the commonest species, when the real forest was compared to the pretend forests there were more than the expected number of recruits in places where parental trees were sparse, and fewer in places where the parental trees were plentiful. Even some rare

Figure 81 This sun-filled commercial teak forest in Sabah's Danum Valley is a dramatically simplified ecosystem compared to the rainforest that it replaces. Could natural selection eventually drive the inhabitants of such a simplified ecosystem to become more different from each other, restoring something like the diversity of the original forest?

species showed the same pattern, though the statistical significance was less because the numbers were small.

Thus, the data fit the expectations of the Janzen–Connell and the N-C models. There really is frequency-dependent recruitment going on in the Barro Colorado Island rainforest plot.

I applied this pretend forest method to other forest plots in different parts of the world, and found the same pattern. One of these was the astoundingly diverse forest at Lambir in the Sarawak province of northern Borneo, where there are over 1,400 different species of tree in a half-square kilometer plot.

Others have since replicated these results. For example, H. S. Dattaraja and his colleagues have extended them to the dry tropical forest at Mudumalai that we met at the beginning of this book, and found the same frequency-dependent pattern.

New puzzles about diversity

It seems clear that frequency-dependence is an important mechanism for maintaining diversity in these tropical forests. But, as with every answer in science, the number of questions that the answer raises goes up exponentially. These questions bear directly on what will happen to life's diversity in the future.

One question is a purely scientific one. What phenomena are actually responsible for the frequency-dependence that we see? Does it, as Janzen and Connell suggested, result from interactions between the trees and an "invisible world" made up of bacterial and fungal pathogens and a swarming collection of insect and animal herbivores? The other major model for the maintenance of diversity, the niche-complementarity hypothesis, also requires selective mortality, but in this model deaths result from competition for limited physical resources. Which of these two models predominates in the rainforests? We do not yet know, but I suspect that each will be found to play a role, and that Janzen–Connell will turn out to be more important for some species and N-C for others.

Janzen–Connell and N-C both predict that to maintain the health and diversity of an ecosystem there must be selective mortality, in which some members of a species are more likely to die than others. Just as such differential mortality is essential for the evolution of a species, differential mortality is essential to maintain the health and diversity of an entire ecosystem.

The burial service in the Book of Common Prayer says, gloomily: "In the midst of life we are in death." I prefer to think that, in the biological world, in the midst of death we are in life.

Second, we can ask whether all the frequency-dependent selection that we see is actually maintaining diversity. The answer to this question seems to be yes.

My colleagues and I measured the diversity of several forest plots from around the world's tropics. We found that the commoner species in each part of the forests died off at a faster rate than the rarer ones. This process maintained diversity, not by adding species, but by ensuring that species that happened to be locally rare did not disappear.[9]

In an ideal world, diversity might actually increase as ecological niches multiply and new species evolve that can take advantage of them. But this is not an ideal world. We are destroying and fragmenting forests and other ecosystems. The animals and birds that help to disperse seeds are being hunted down. Even the insects that contribute to the delicate frequency-dependent balance in the forests are being reduced in numbers as older trees, with their thousands of ecological niches, are replaced by scrubby second-growth. And as our climate changes, fire is sweeping through forests that have never before experienced it.

All these changes put at risk the specialized pathogens and herbivores that abound in these complex ecosystems. One might be less inclined to mourn their loss than the disappearance of the many colorful and charismatic creatures that also populate the forests. But all these nastier and less lovable organisms turn out to be essential for the survival of the charismatic organisms as well. If the deadly invisible world is lost, or if its own diversity is diminished, then the motive force that drives the Janzen–Connell effect will disappear and the diversity of the world's forests will be reduced.

Darwinian evolution and the evolution of ecosystems

Ecologists deal with the measurement of present-day diversity. But we have already seen that rainforests have changed greatly since the beginning of the Age of Mammals. Large numbers of new species have appeared through the evolutionary processes that we have explored in this book. And it turns out that ecology can work hand-in-glove with evolution to increase diversity.

We saw in the previous chapter that during the process of speciation new species will be actively driven apart from each other genetically if any hybrids between them cannot occupy any of the available ecological niches. The two species will become more and more different from each other, and therefore less likely to mate and produce hybrids. The fitness of members of both species will be harmed if they succumb to the allure of the exotic

Figure 82 (*opposite*) These stingless bees in Sarawak's Lambir Forest defend their nest, and their host tree, fiercely. Although the bees cannot sting, they are very good at biting!

species next door and waste their genes by copulating with them. There will be selection for behavioral and chemical reproductive isolating mechanisms that make those alien creatures less desirable.

But once two species are well and truly driven apart by selection for such isolating mechanisms, any further selection for these mechanisms diminishes and eventually disappears. Now the two species are so different that they no longer attract each other. Does their subsequent evolution now slow to a crawl, governed only by the gradual accidental accumulation of further mutational differences over long periods of time? Or do selective pressures exist that can continue to drive the species even further apart, making them more different from each other and increasing the overall diversity of the ecosystem in which they live?

If the two species come to occupy clearly different ecological niches, the answer is yes. Natural selection will continue to drive their divergence as they adapt to the different niches. But in a rainforest the ecological niches are crowded and overlapping. The further divergence of two species is likely to be a slow process, especially if they live near each other so that their habitats remain similar. Once speciation has happened, the further accumulation of mutations that make the species more and more different should be as slow as watching paint dry.

But here is where the evolution of entire ecosystems comes into play. Interactions with the entire ecosystem can speed up species divergence. Tree species can become extremely different from each other even if they have similar light and water and nutrient requirements.

Consider two species of plant in a forest that had a recent common ancestor. Because they are so closely related they are still rather similar in their appearance, with similar biochemistries, stem structures, leaf shapes, and so on.

Why have they become different species? There are many possible reasons. For example, pollinators may play a role, as has been shown in beautiful experiments by Doug Schemske of Michigan State University and his colleagues.

Schemske studied two species of monkey-flowers that live in overlapping habitats on the slopes of California's Sierra Nevada.[10] The two species differ in the shape and color of their flowers. As a result they attract different pollinators—one species attracts native bees and the other attracts hummingbirds.

Almost certainly, these two species arise from their recent common ancestor because different populations of the ancestral species lived in places where only one of the two pollinators was prevalent. The two populations were then selected to attract the pollinator that was commonest in their region. Now, because the two species are visited by different pollinators, they can no longer exchange genes even when they live in the same meadow.

Schemske found that although the two species are effectively isolated from each other, gene flow is still possible between them. Hybrids can be readily produced when the experimenter takes over the role of pollinator, and deliberately paints the stigmas of one species with pollen from the other. The resulting hybrids are fully fertile, though they are a little less fit than the original species.

When Schemske mated these hybrid flowers with each other, he produced a wide variety of new flower types with mixed-up characteristics. Some of them do not attract either pollinator. This means that in nature, any accidental hybrids would be at a disadvantage. This disadvantage has helped to drive the two species towards their different pollinators. When he tracked the genes of hybrid flowers that attracted one, both or neither pollinator, Schemske was able to identify some of them. These genes influence flower color and the amount of nectar produced. Bees prefer more orange-colored flowers, and hummingbirds prefer flowers that make copious amounts of nectar. These results provide a beautiful example of how speciation actually happens in the real world.

So, returning to our two forest plant species, let us suppose that they have speciated by some similar process. Like Schemske's monkeyflowers they can no longer exchange genes in their native environment, even when they happen to grow next to each other. Now what? The disadvantage of hybridization has helped to drive the forest trees to become two different species, but that disadvantage has now disappeared because hybridization no longer happens.

Here is where the entire ecosystem starts to play a role in driving the species further apart. If such closely related species are going to thrive in their ecosystem, they must begin by simply surviving. They can do so in many ways. One way is to hitch a ride on the Janzen–Connell or the N-C bandwagons.

The two species are still so similar that they are preyed on by the same set of parasites and herbivores. When both species are common in a certain part of the forest, their progeny are equally selected against by the enemies that they share. When both species are rare, their progeny can survive more readily because they are not preyed on as much.

So far as the pathogens and herbivores are concerned, the two species are equivalent. But there will immediately be an advantage to any members of both species that happen to be different, because they will be able to escape some of the pathogens and herbivores that prey on them.

Host species are always evolving new defenses, and pathogens and predators are always evolving new ways to overcome them. But in the case of our two species there is a new dimension to this continuing conflict. If one of the species can escape at least some of the pathogens that prey on it, it will be able to move into parts of the forest where the other species cannot thrive. This will enable it to take advantage of niche-complementarity. It will now be able to utilize environmental resources that would otherwise have to be shared with the other species.

In short, it is a win-win situation for the two species. If they continue to move apart on many fronts—defense mechanisms, biochemistry, physiology, and so on—they will be preyed on by fewer predators and parasites. And because each of the host species is able to survive in parts of the forest infested with pathogens that no longer harm them, they will have access to physical resources denied to the other species.

Of course, this advantage is only temporary. Soon some of the predators and parasites will evolve to overcome the tree species' fresh defenses. Repeated rounds of ever-evolving defenses and attacks will continue to drive the two species apart, long after they have lost any possibility of hybridizing with each other. At the same time, if they accidentally come to share predators and pathogens with other species in the forest, they will be driven away from those species as well. This process contributes to the astounding diversity that dazzles us when we visit complex ecosystems such as rainforests and coral reefs.

As the host species are driven further and further apart by these selective pressures, the parasites and herbivores that prey on them must also become more diverse. The result is an ecosystem with such a diverse set of host species that no single pathogen or herbivore is versatile enough to attack every host.

The preservation of diversity, at the level of both the visible and the invisible worlds, is essential if ecosystems are to remain healthy and balanced. Trees and bushes, pollinating insects and birds, seed-eating pigs and browsing tapirs, frugivorous bats, birds and primates, the fierce ants and the stingless bees that defend their host trees, beetles and bugs in unimaginable profusion, fungi that can benefit or damage their hosts, and the myriad bacteria and viruses that form the invisible base of this complex edifice—all are linked in a fragile web of life and death. It is these dynamic interactions that make every encounter with a complex ecosystem such an adventure for an alert Darwinian tourist.

Ecology, malaria, and the origin of humans

Up to this point we have been talking about trees in the rainforest, but of course the ecological interactions that encourage the maintenance and evolution of rainforest diversity must be operating in the rest of the living world as well. And that includes the evolution of our own species. Like the trees of the tropical rainforests, we have evolved to fight off our pathogens. And, like the tree species, we have been driven apart genetically from our nearest relatives as we have succeeded in keeping our shared pathogens at bay.

My colleague Ajit Varki at the University of California San Diego has been studying the unique evolution of our species for much of his scientific career. Twelve years ago he and I set out to quantify a set of anecdotal observations. When we compare ourselves with our nearest living relatives, the great apes, we seem to be extremely different in our appearance and behavior. We decided to build an evolutionary tree in order to visualize this relationship.

To build the tree we used a wide variety of physical and behavioral characters that we share with chimpanzees, bonobos, gorillas, and orangutans. These included the ability to walk bipedally, the difference in size between the left and right hemispheres of the brain, the amount of thumb mobility, the length of time spent nursing the young, and several other characters.

The great apes in the tree grouped fairly closely together, but humans stood out dramatically. We occupied the end of a long branch that stuck out from other, shorter branches like a fishing pole. This tree-building exercise showed clearly that we really are dramatically different from even our close relatives.

Many of these differences between ourselves and the apes can be attributed to selective pressures that have increased our intelligence and toolmaking ability, and that have encouraged the development of complex social interactions within human groups. As our culture has become more complex with time, we have evolved to take advantage of the opportunities that have been provided by our culture. The result of this evolutionary feedback loop, involving our brains, our bodies, and our culture, has been a uniquely intelligent, uniquely dexterous, and uniquely social species.

But, remarkable as our evolution has been, there is no evidence that our uniqueness has been caused by anything other than the basic evolutionary processes we talked about in Chapter 2: mutations, natural selection, chance events, and the genetic reshuffling of our gene pool that happens every generation as a result of sexual reproduction. Darwin always insisted that our origins can be traced to the same evolutionary processes that have led to the diversity of the rest of the natural world, and he was right.

Some of the evolutionary changes that have contributed to our emergence as a separate species have nothing to do with our culture, but everything to do with our resistance to disease. In 1998 Ajit Varki's group and another group of scientists discovered the genetic basis for a molecular difference between ourselves and the apes.[11] This difference, found on the surfaces of the cells of our bodies, has driven us dramatically apart from our closest ape relatives. To understand the nature of this difference, we must delve into the submicroscopic world of molecules.

Each of the 100 trillion cells in the human body is surrounded by a delicate cell membrane that shields it from harmful molecules while allowing essential ones to pass through. The outer surface of this membrane, which directly confronts the environment, is studded with a maze of treelike chains of sugar molecules. These chains branch and join together to form a lattice-

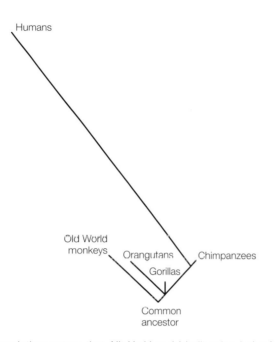

Figure 83 The evolutionary tree that Ajit Varki and I built using behavioral and physical characteristics of the humans and great apes. We humans are clearly different from our close relatives. We stick out from the tree like a fishing pole. (From C. Wills, *Children of Prometheus* (Cambridge, MA: Perseus Books, 1998), 200.) © Christopher Wills.

work, providing a first line of defense against the invasion of bacteria and viruses.

The sugar molecule chains are attached to proteins and fatty molecules embedded in the cell's membrane. These embedded molecules are free to slither around within the two-dimensional world of the membrane like skaters on a skating rink. They drag the sugars with them, forming ever-shifting *cheveaux-de frise* of chemical defenses against invaders.

The outer tips of most of the chains of sugar molecules are occupied by modified sugars called sialic acids. The predominant form of sialic acid in most mammals has a teeny negatively charged arm that stretches out to meet the environment. These charged sialic acids interact readily with positively charged molecules outside the cell.

The equivalent sialic acid molecules in humans, however, do not have this negative charge. The tips of their chains of sugars are uncharged. And this, as Ajit discovered, is because we have lost the enzyme that normally adds the extra charge. Although all our ape relatives have an intact charge-adding gene, the form of the gene that all humans carry has been devastated by a DNA deletion that effectively destroys its function.

This mutational change spread through our remote ancestors, replacing the allele that coded for the functional enzyme. The change made the surfaces of the cells of those ancestors different from those of all their close relatives. When did the mutation happen? DNA evidence suggests that it took place about two million years ago, which puts it back at around the time when ancestors of ours called *Homo erectus* first left Africa for Asia.

Biochemical evidence now reinforces this genetically derived date. Less than a million years ago, our lineage diverged from the ancestors of the Neanderthals, who left Africa to populate Europe and the Near East. Neanderthals are now extinct, but through a chemical tour de force Ajit and his colleagues were able to isolate sialic acid from Neanderthal bones and show that it was chemically identical to that carried by modern humans. This independent evidence shows that the mutational change must have happened before the human–Neanderthal split, which puts it at a minimum of about a million years ago.

What drove this change? Ajit and his colleagues note that the most dangerous human malaria parasite, *Plasmodium falciparum*, is able to attach itself firmly to the less-charged form of sialic acid that is found on the surface of human red blood cells.[12] This attachment helps it to invade our cells. The best-studied chimpanzee malaria parasite, *Plasmodium reichenowi*, attaches itself to the more charged form of sialic acid that is found on chimpanzee red blood cells. The human parasite cannot attach to and invade chimpanzee red blood cells, and vice versa.

Other pathogens that cause human and ape diseases also bind to sialic acids, but malaria is by far the deadliest. Malaria parasites account for more than three million deaths annually, and many more millions of people are left severely debilitated. The relative impact of this disease must have been much greater on the smaller human populations of the past.

Our ancestors began to diverge from those of chimpanzees six million years ago, in Africa. For four of those six million years they continued to be susceptible to the chimpanzee malaria parasite, *P. reichenowi*.

Thus, when the less-charged mutant sialic acid appeared in our ancestors about two million years ago, it was a big deal indeed. It was strongly selected for, because it immediately conferred resistance to this dangerous malaria. Our lucky ancestors were not totally freed from malaria, because there are three other less dangerous malaria parasites that also prey on us. Nonetheless, at one stroke, they had rid themselves of the most debilitating form of this disease.

We do not know the full benefits that followed from the emergence of this new immunity to *reichenowi* malaria, but it might have helped our African ancestors survive difficult times. They may have been given some breathing room as they continued on their new evolutionary path towards more sophisticated tool use and more complex social structures. It may not be a coincidence that the first stone tools, the first human-like hands, and the first signs of organized hunting all appeared in Africa at about the same time as the sialic acid mutation. Many other factors certainly contributed to these essential events in our history as a species, but freedom from severe malaria must have been a big help.

All good things come to an end, and at some point the inevitable happened. A mutant strain of *P. reichenowi* arose that could attach itself to the less-charged sialic acid. This mutant strain eventually gave rise to the most dangerous human malaria, *P. falciparum*. The *falciparum* DNA family tree is a branch of the *reichenowi* tree, but so far it is difficult to say when it branched off. It may have been as recently as 10,000 years ago, or as long ago as a million.

The mutation that allowed *falciparum* to attach itself to our ancestors' red cells occurred in an invasion receptor gene of the parasite that is called EBA 175. When the mutant strain of malaria appeared it exploded in sub-Saharan Africa and spread to the Middle East, but not to temperate regions because the parasite cannot survive long winters. The result has been, as Jared Diamond has pointed out in *Guns, Germs and Steel*, that societies in tropical areas that suffer from malaria have been held back relative to those of less malaria-ridden regions like Europe.

This story is far from complete. In humans, many genes are involved in resistance to parasites and to bacterial and viral diseases. Ajit Varki and others have found mutant forms of some of these genes in our species. These mutant forms separate us even further genetically from the apes. Perhaps the spread of these mutant genes has also been driven by their ability to protect us from the diseases that we used to share with our close relatives.

This story of humans and malaria resembles the one I proposed earlier for our hypothetical pair of plant species in the forest. I pointed out that the two plant species would still continue to be susceptible to a shared set of parasites as they begin to speciate and move apart genetically. Just as has happened with humans and malaria, they would then be selected for resistance to the parasites and pathogens of the other species. And the parasites and pathogens would evolve in turn, driving the plant species even further apart.

As we examine the underlying processes of evolution, the same themes emerge again and again. Earlier in this chapter I pointed out that the pattern of recruitment and survival in rainforests around the world shows how the diversity of the rainforest itself may have evolved. I suggested that the diverse tree species of the rainforest may have been driven apart genetically by two factors. The tree species were able to avoid the pathogens and predators that they shared with closely related species, and they were able to invade regions of the forest that had previously been closed to them.

Now, evidence is emerging that humans have been subjected to selective pressures similar to those of our hypothetical plant example. We escaped some of our most dangerous parasites in Africa by acquiring a sialic acid that was different from the one belonging to our chimpanzee relatives. Perhaps our ability to escape from our shared parasite gave us the chance to acquire the technology and cultural cohesion necessary for us to escape from Africa itself.

At first blush there might seem to be a huge gap between the origins of the ecological and genetic diversity of rainforests and the origins of the differences between our own species and our ape relatives. But as we learn more about the evolutionary forces involved, a common theme emerges. In the book's second part, we will look further at how our own history has been shaped by such evolutionary pressures. In the process we will discover additional connections between ourselves and the rest of the living world.

PART II
The Human Story

This Hadza hunter-gatherer near Lake Eyasi in Tanzania has found and opened a wild bee nest, using keen observation coupled with tool-making skills. In this second part of the book we explore how we and our close relatives have evolved the many remarkable abilities that define our humanity.

6

How Domesticated
Animals Changed
the World

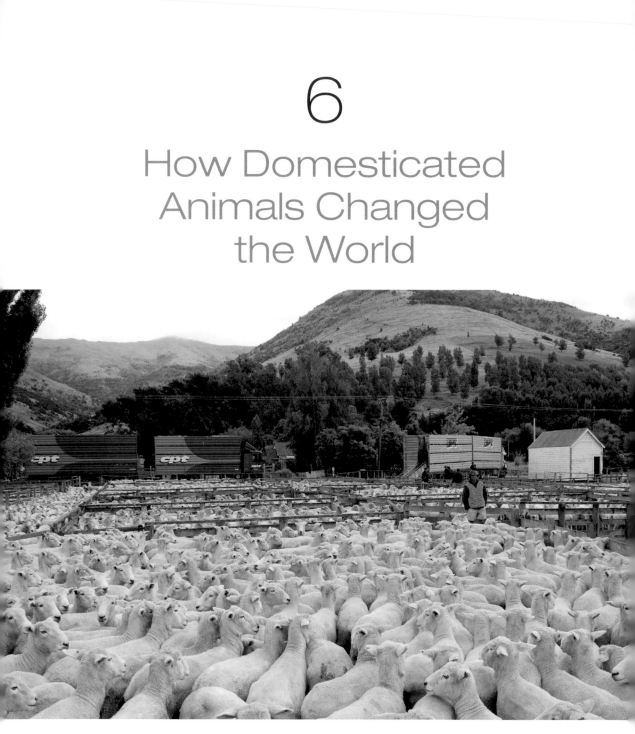

Figure 84 These freshly sheared sheep on New Zealand's South Island await transport back to their farms. The landscape of New Zealand, which was empty of mammals other than bats before the arrival of humans, has been largely transformed by the introduction of domestic animals.

On August 1, 2008, the sun rose into an absolutely cloudless sky over Mongolia's Altai Mountains. The stark and treeless mountain slopes, and the wide desiccated valley that lay to their west, soon began to shimmer with heat haze as the temperature climbed. On the western horizon a line of hills marked the boundary between western Mongolia and China's Xinjiang province.

My companions and I were extremely happy to see the cloudless sky, because that afternoon there would be a total solar eclipse. Nothing ruins the big day for an ardent eclipse watcher more than clouds or haze. Up until the moment that the eclipse actually begins there is no way to predict with certainty what the fickle atmosphere will do. The weather had been unsettled during most of our long journey to this remote part of central Asia, and it had rained the night before. So we greeted the flawless dawn with excited relief, and hoped that the weather would hold.

Soon after sunrise we drove out to the west, away from the Altais and toward the Chinese border, to hunt for some petroglyphs at the head of a dry canyon. On the slopes of a mountain range to the north, still blue in the morning shadow, we glimpsed a small herd of black-tailed gazelles. They fled as soon as they saw our Land Cruiser—not surprisingly, since they are so intensively hunted.

The steppes of western Mongolia were once home to vast herds of gazelles and wild horses. The mountains that loomed above the steppes, far greener in the past than they are today, supported thousands of ibex and giant Altai Argali sheep. Eight hundred years ago Genghis Khan and his tribesmen took part in vast hunts that swept across both steppes and uplands, driving the animals before them for 1,000 kilometers. At the end of the drives the animals were herded into a huge panicked mass. Then Genghis and the other leaders, in order of rank, slaughtered them for days on end until they had to stop from exhaustion. The remaining animals were left to be killed by the ordinary tribesmen.

It must have seemed to the Great Khan and his followers that this immense natural resource was inexhaustible. Alas, it was not. Now the central Asian gazelles, ibex, and other animals are perilously close to extinction. Hunters pay large fees to kill the few that remain.

Figure 85 These Bronze Age petroglyphs in the Altai Mountains of western Mongolia, some of which are 2,000 years old, provide a glimpse of a time when wild animals roamed the Central Asian steppes in uncounted numbers.

Some of the Bronze Age petroglyphs that we found in the canyon dated to about 2,000 years ago, 1,200 years before the days of Genghis Khan. They illustrated a great profusion of grazing animals, giving us a glimpse of a time when animals must have been even more abundant than when the Great Khan brought the nomadic tribes together under one leader.

Wild herds began to decline in numbers as a result of human hunting long before the time of Genghis Khan. The shortage of wild animals provided an impetus to domestication. This behavioral transition from hunting to herding was one of the most important in human history, a change from

188

simply living off the land to the acquisition of a far more predictable source of meat and textiles and entirely new food resources like milk and cheese.

There were also evolutionary changes in the people who domesticated the animals. These changes were driven, not directly by the domestication of animals and plants, but by the resulting alterations to the ecological niches that we occupy throughout the planet. As we saw in the previous chapter, recent evolutionary changes in our species have been driven by a feedback loop involving our environment, our genes, our bodies, and our behaviors. The domestication of animals dramatically altered all aspects of the human environment, irreversibly changing important components of this feedback loop. We will see, in this and the following chapters, how such feedback processes have shaped ourselves and our world.

How dogs were domesticated

As we returned from the petroglyph canyon to await the eclipse, we stumbled into a time warp that took us back to one of the most important of those early domestication events.

Our guide Tugso Tugso got word that a boy in one of the villages had found a wolf cub. We detoured to the tiny village of Oench, a cluster of mud-walled houses scattered along a dusty riverbed that cut across the truck route from the Chinese border. There the boy proudly showed off the little six-week-old cub that he had found abandoned in its mother's lair. There was no sign of what had happened to the mother.

The cub belonged to the wolf subspecies *Canis lupus chanco*, which is distributed across Asia from Kazakhstan to Korea. His eyes were brown rather than the intense yellow that forms such a striking feature of North American wolves. He was determinedly gnawing on a bone, and from his possessive attitude it would have been extremely unwise for us to offer him a friendly finger.

What would happen to this cub? We did not know, but other cubs that had been found by the children of the village over the years were raised for a short while and then released when they were no longer puppies.

At some point in the distant past, a similar encounter between a central Asian nomad boy and a wild wolf cub may have been the starting point in the long journey from wolf to dog. There is some evidence that in eastern Asia, probably between 15,000 and 25,000 years ago, dogs were domesticated from wolves. In the dusty street of that little Mongolian village we had glimpsed an echo of that time, when many groups of nomadic peoples were moving across the green slopes of the Altai Mountains and when such captures of wolf cubs by curious children must have happened repeatedly.

The children must have tried to raise the cute little animals, but as most of the wolf cubs grew larger they bit too many fingers and were banished from the household. Occasionally a wolf cub was found that was especially friendly and could join the family. It soon paid its way, acting as a guard and raising the alarm at night when it heard or smelled animal or human intruders. Perhaps it killed rats in its spare time. And eventually its descendants learned to herd and guard the other animals that these nomads began to domesticate. It is likely that the tamed wolves of the encampments served as a source of meat during grim times, as they have done more recently for Eskimos.

Wolves must have been difficult to tame at first. Most wolves do not relate readily to people, a fact that was quantified by Brian Hare and his colleagues at Harvard and Leipzig.[1] In their experiments they confronted dogs and wolves with a pair of boxes. Both boxes smelled of food, but only one of the boxes had food hidden in it. The dog or wolf had to find which box was the right one.

In each trial a human whom the animal had never met was also present in the room. Dogs immediately looked to the human for guidance, and responded readily when the human touched or pointed to the box with the food. They responded correctly even when the human merely glanced at the box.

Most of the wolves sniffed around the boxes at random, ignoring the humans and their signals. But a few of them did respond correctly at greater than chance levels when the human both looked at the box and touched it. These wolves may simply have been more accustomed to humans, but it is also likely that there is variation in wolf behavior. As a result of such variation some of those East Asian wolves from 15,000 years ago might have been easier to tame than others.

Figure 86 A wolf cub, abandoned by his mother in a den near the western Mongolian village of Oench, was brought to the village by a young boy. The cub is so cute that he cries out to be domesticated.

As I emphasized in Chapter 2, natural selection can only produce genetic changes in populations if there is genetic variation available that the selection can act on. The same requirement applies to artificial selection. If there were no heritable differences among the behaviors of the wild wolf cubs, then the people of the encampments could have gone on selecting the tamest cubs for centuries, but their selection would have had no effect. The cubs' babies would have been as wild as the wolves that still roamed free. Variation among the genes that contribute to the behavior of the selected animals is essential if artificial selection for domestication is to succeed.

Even before the advent of DNA sequencing, biologists had suspected that domestic dogs originated from *C. lupus chanco*, the same subspecies as the cub that the boy in the village had found. The lower jaws of domestic dogs share with the *chanco* subspecies an oddly shaped hook at the top of the ascending part of the jaw. The hook is not found in the jawbones of other wolf subspecies.

But such evidence is highly circumstantial—dogs could have evolved their bony hook independently of the Asiatic wolves. We must turn to DNA evidence to get a clearer idea of where dogs came from. Mitochondrial chromosomes provide the information we need.

Mitochondrial DNA—an introduction

Because mitochondria and their DNA play a central role in our understanding of how animal domestication happened, and of how our own species evolved, it makes sense to pause for a moment to introduce the friendly little creatures.

Mitochondria are found in most of our cells, and in most of the cells of wolves and dogs. These tiny structures, bounded by two concentric layers of protein-studded membrane, are all that remain of bacteria that were formerly free-living. Almost two billion years ago some of those free-living bacteria found their way inside our remote single-celled ancestors. Perhaps they were eaten by those teeny ancestors of ours, or perhaps they simply bullied their way into our ancestors' cells just as malaria parasites and other organisms can invade our cells today. And it is quite possible that the first wave of those bacterial invaders, like today's intracellular parasites, made our ancestors sick.

It turned out, however, that being invaded by these bacteria yielded a fantastic evolutionary payoff for our distant ancestors. The payoff was so huge that it greatly outweighed any damage that the invading bacteria might have caused. This is because our ancestors had previously, in a fit of carelessness, lost their ability to use oxygen for respiration. This was a really dumb thing to do, because organisms that can use oxygen can squeeze eighteen times as much energy from a sugar molecule as those that cannot. Luckily for us, the invasion of our cells by the bacteria gave our ancestors back that ability.

Why were our ancestors so careless as to lose the ability to burn food with the aid of oxygen in the first place? Perhaps there was an evolutionary trade-off involved. In order to respire efficiently and gain the maximum from their food, single-celled organisms need special cell membranes that can generate a strong electric charge difference between their inner and outer surfaces. If there are breaks in the membranes they will lose the charge difference, just as batteries do when they short out. These cells are unable to wrap themselves around and engulf solid pieces of food without making breaks in their membranes thus destroying this essential charge difference.

So it may be that our ancestors made a Faustian bargain: they lost the ability to burn food using oxygen, but in exchange they were now able to modify their membranes so that they could engulf smaller organisms like tiny bacteria and algae.

Although they were now able to eat other creatures, becoming the world's first predators, they could not extract much food energy from their victims. Luckily, this downside to their Faustian bargain was short-lived. When the bacteria that would eventually become mitochondria invaded the cells of our carnivorous ancestors, their ability to carry out aerobic respiration was restored. And our ancestors did not have to lose the ability to eat solid food, because all of this beneficial respiration was now taking place deep inside their cells instead of on the surface. The way was open for our ancestors to become highly effective predators. They were able not only to catch other creatures but also to extract lots of energy from them. Our ancestors became the tigers of that simple world.

The respiring bacteria soon became permanent fixtures in their new hosts, entering into a relationship with them called *endosymbiosis*. Our ancestors nurtured the bacteria carefully, passing them like family jewels from one generation of host cells to the next.

The mitochondrial descendants of those bacteria now live in immense numbers in most of the cells of our bodies. They are cute little ovoid structures, still looking a bit like bacteria. As they pump energy-rich compounds into our systems, they keep our metabolisms running at full blast.

After two billion years the chromosomes that the ancestors of the mitochondria brought with them have lost most of their genes. Some of their genes have been inserted into our own chromosomes, while others were probably unneeded and are lost forever. Only thirty-four genes now remain on the little ring-shaped piece of bacterial DNA that our mitochondria still carry.

We inherit mitochondria and their chromosomes only from our mothers. Our fathers have mitochondria too, because otherwise they would be dead, but the mitochondria carried by the sperm are destroyed when the sperm enters the egg. Only the females' mitochondrial chromosomes are passed down from one generation to the next.

A brief history of dogs

This capsule history of the mitochondria now leads us into how we can probe the histories of dogs and other domesticated animals.

Most of the genes on our chromosomes are mixed up through genetic recombination and inherited from both parents, so that our children are always different from their parents. But our little snippets of mitochondrial DNA are exactly like those of our mothers. The same pattern of inheritance applies to dogs, which inherit their mitochondrial DNA in exactly the same way. But over time, mutations will accumulate in mitochondrial chromosomes, and they do so relatively rapidly. As a result of these accumulating mutations, dog mitochondrial chromosomes differ a little bit from those of their wolf ancestors but can still be traced back to them.

As we saw in the first chapter, comparisons of DNA sequences can help to sort out the relationships among any group of living creatures. The more closely related the DNA sequences of two organisms are, the closer they tend to be on the family tree. Relying on these relationships, Peter Savolainen of Stockholm's Royal Institute of Technology has built a family tree of mitochondrial DNA sequences from dogs and wolves around the world.[2]

The tree that Savolainen and his colleagues obtained was a bit difficult to interpret, because there had only been time for a small number of mutational differences to accumulate in the 600 base pairs of dog and wolf DNA that they compared. But two things were clear. First, among all the Old World and New World wolves, it is the Asiatic wolves that are the closest relatives of all domestic dogs and therefore the most likely to be their ancestors. The anatomists who had fingered *C. lupus chanco* as the dog ancestor were right. The little wolf cub of Oench Village comes from the same group of wolves as the ancestor of all dogs.

Second, they found that dog populations from eastern Asia seemed to harbor the most genetic variation. This is the pattern that would be expected if dogs had originated in East Asia and had later given rise to subpopulations elsewhere. The original population would have brought a good deal of genetic variation from their wolf ancestors, but because later subpopulations are subsets of the original population they would be expected to carry smaller amounts of genetic variation.

Both of these findings supported an East Asian origin for at least some dogs. But was the origin single or multiple? Were dogs domesticated many times, or only a few times—perhaps only once? And when did domestication happen? Here the data are more equivocal.

Different dog DNA sequences have been found to fall within different branches of the Asiatic wolf tree. This could be because dogs were domesticated more than once from different female wolves—the data suggest that domestications could have happened at least six different times. But these six events did not give rise to different kinds of dogs—each of the many present-day dog breeds carries several of these wolf lineages.

The number of mutations that have accumulated in dog mitochondrial chromosomes allowed Savolainen to estimate that the multiple origins of dogs took place early, probably at least 15,000 years ago and perhaps twice as long ago. This is thousands of years before the domestication of other animals such as the horse. Dogs may have been domesticated independently in different parts of East Asia around that time, or there may have been a cluster of domestication events involving several wolves. The production by artificial selection of present-day dog breeds took place much later, which explains why each breed carries mitochondrial chromosomes from a number of the ancestral wolf lineages.

Savolainen favored the 15,000 year date, because at that time no dog fossils more than 14,000 years old had been found anywhere in Europe or Asia. But there has been a recent dating of a dog skull in Belgium to 33,000 years ago, which may upend that date and push dog domestication back much further.[3] Alternatively, as we will see later, there may have been several earlier domestications, perhaps including the one in Belgium, that produced tame dog lineages. It is possible these lineages did not survive.*

The immense value of dogs

Dogs were so useful that they began to accompany their masters around the world soon after they were domesticated. DNA sequencing and carbon-14 dating from New World dog burials show that dogs accompanied the first humans to the New World, perhaps 14,000 years ago. If Savolainen's date is correct, this means that the migration to the Americas may have taken place not long after dogs were first domesticated in Asia. The people who traveled across the Bering Strait may have taken some of the very first dogs with them.[4]

Samples of DNA have been obtained from the bones of pre-Columbian dogs from around the Americas. These lineages remained distinct and did not mix with local wolves or coyotes. Right from the beginning of their domestication, dogs tended to live with their human owners and were not tempted by the call of the wild.

But some of humanity's most extensive migrations took place long before the domestication of man's best friend. The very longest was the great migration from Africa across South Asia and into China and Southeast Asia that began perhaps as early as 100,000 years ago and took as much as 50,000 years (see the next chapter). In the final leg of this migration people crossed

---

* A recent paper by vonHoldt et al. (*Nature*, published online on March 17, 2010) provides strong evidence that most dog domestications took place in the Middle East about 12,000 years ago. The paper also presents evidence of some contribution from East Asian wolves, in agreement with the earlier studies. Old dog fossils like the one in Belgium, therefore, may be traces of earlier domestications that have been lost. Remarkably, herding dogs and dogs that depend on sight and smell were selected very early in the domestication process, and seem to have been selected for only once.

a stretch of open ocean from Timor to Australia. They arrived in Australia 50,000 years ago, long before dogs were domesticated.

There were no wolves in Australia for these first migrants to tame. The native wild dog, the dingo, originated in Southeast Asia and came to Australia only 5,000 years ago. The ancestors of the dingoes may have made the journey without human aid, but it is more likely that they were domestic dogs that escaped from the boats of brief visitors to the continent.

Up to that time the Aborigines had never had the companionship and help of dogs. When the dingoes did arrive they were not immediately adopted. Some Aboriginal tribes simply followed the tracks of wild dingoes to help them find game. Others captured pups occasionally and raised them as hunters, but then allowed most of them to escape. Still others raised dingoes for food. A few tribes raised the dogs but used them chiefly to keep warm during cold nights, much as the Incas on the other side of the Pacific used their specially bred hairless orchid dogs.[5] It is possible that a similar wide spectrum of interactions had characterized the original domestication of wolves millennia earlier in East Asia.

Australians survived without dogs for tens of thousands of years, but our ancestors in Africa had to survive without them for millions. The bones of a domesticated dog from the Nile Delta, the oldest yet found on that continent, date from only about 4,700 years ago. As dogs spread south from Egypt they were rapidly adopted by different tribes, resulting in unique breeds like the Khoi dog of the San peoples and the strangely barkless Basenji.

A recent study by Carlos Bustamante at Cornell and his colleagues finds that African village dogs have as high a level of genetic variation as East Asian dogs, calling into question Savolainen's argument that the variable dog populations of East Asia mark the origin of domestication.[6] But in order for there to have been a separate African domestication of dogs, East Asian wolves would somehow have had to make it to Africa, or to have been brought there. Argument continues, but I suspect that the Asian origin of domestic dogs will continue to hold up.

It may be lucky for African ecosystems that dogs arrived there so recently. If hunting tribes had been aided by dogs, they might have done irreparable damage to their environment.

Humans have had a dramatic impact on their environment for a long time. The earliest Americans, like the earliest migrants to Madagascar and New Zealand, seem to have killed off most of the large animals that they found in their new homelands. These human-caused extinctions, some better documented than others, have been termed the "Pleistocene overkill" by Paul Martin of the University of Arizona.[7]

There are many questions about the role of humans in the widespread extinctions of horses, camels, mastodons, and saber tooth cats that took place 10,000 years ago in North America. The evidence for human overkill in mastodon extinctions is strong and growing.[8] But was human over-hunting entirely responsible, or did the changing climate at the end of the last Ice Age play a role? And did dogs, which we know accompanied the first human migrants to the New World, help their human masters to drive so many species to extinction?

The evidence is much stronger that humans killed off the giant flightless birds of Madagascar and New Zealand. We also know that dogs accompanied the first Maori settlers to New Zealand seven hundred years ago. The first European explorers saw these *kuri* dogs being used to hunt small birds, and as a source of food.[9] But we do not know whether dogs also helped the Maori to hunt the giant flightless moas, the last of which disappeared not long before Captain Cook landed there in 1769.

The Maori dogs themselves disappeared soon after Cook's arrival, replaced by more useful European breeds.

On a personal note, having tramped through some of the steep, thickly forested slopes of New Zealand's South Island, I cannot see how the Maori could possibly have hunted down the last moas in those remote regions without the aid of tracking dogs.

Loren Eiseley and others have questioned whether humans were really responsible for many cases of apparent Pleistocene overkill. And why, they ask, did most of the large animals of Africa survive even though our ancestors hunted them for millions of years? Paul Martin responded by pointing out that extinction rates in Africa actually did increase at the same time as human influence was spreading. About forty percent of the large animal genera in Africa have disappeared in the two and a half million years since

Figure 87 A jackal, *Canis aureus*, at Ranthambhore, searching for prey. As the last of the light faded, the herd of deer that the jackal and its companions were hunting fled past me into the shadows.

the beginning of the Pleistocene, the same period of time during which our ancestors were learning to hunt them.

It could be that what saved Africa from the devastating overkills of California, New Zealand, and Madagascar was simply the mind-boggling abundance and diversity of the animals that roamed the African plains and forests. Human populations were also small throughout much of the Pleistocene, and people could not live in some parts of the continent because of malaria and sleeping sickness. But one wonders what damage our ancestors might have done, despite these constraints, if there had been hunting dogs to help them.

Luckily for Africa's surviving diversity, the wild doglike animals living there appear not to be easily tamable. The wild dogs of southern and eastern Africa, *Lycaon pictus*, are separated from the Eurasian wolf by at least three million years of divergent evolutionary history. These wild dogs are experts at hunting animals such as impala. They often chase their prey in relays, while other members of the pack peel to the left and right and execute pincer movements to block attempts at escape.

If *Lycaon* wild dogs were ever tamed by tribes in south and east Africa, we have no record of it. While they could in theory be tamed, the results might have been hard to live with. The anal glands of *Lycaon*, like those of hyenas, produce a sickening odor.

The same puzzle arises with the wild dogs of India. Reddish wild dogs (dhole, *Cuon alpinus*), remote relatives of the African wild dogs, are still found in India, though only about 2,500 remain on the subcontinent. I glimpsed a pack of them near Mudumalai in Tamil Nadu, but they fled before I could photograph them. These dogs, too, seem not to have been domesticated, though why they should not have been remains a puzzle.

Brian Houghton Hodgson, an English civil servant and amateur naturalist who served in northern India during the first half of the nineteenth century, succeeded in raising a dhole cub and found it responsive to his training. An account of his efforts in the *Proceedings of the Zoological Society of London* for 1833 (p. 121) says:

Adults in captivity made no approach towards domestication; but a young one, which Mr. Hodgson obtained when it was not more than a month old, became sensible to caresses, distinguished the dogs of its own kennel from others, as well as its keeper from strangers; and in its own conduct manifested to the full as much intelligence as any of [Hodgson's] sporting dogs of the same age.

But Hodgson does not say whether the puppy began to smell noxious as it grew into an adult, or what its ultimate fate was.

Dholes might have been difficult to live with, but surely Indian wolves are a different matter.[10] Because India was on the early migration path from Africa to Southeast Asia and Australia, modern humans have occupied the southern part of that subcontinent for at least 50,000 years. And wolves were plentiful in India. Why were wolves not domesticated there as well as in East Asia?

For most of that 50,000 years, contacts between humans and the wild world of the forests and jungles must have been part of everyday life. Stories

Figure 88 I encountered this handsome chital stag (*Axis axis*) the next day, under conditions more favorable for photography. Hunting scenes like the one I had glimpsed are becoming increasingly rare as India's wildlife disappears.

of remarkable interactions between wolves and Indian villagers abound. We all remember Rudyard Kipling's *Jungle Books*, his collection of stories about Mowgli, the little Indian boy who was raised by wolves. (Or at least we have been exposed to the saccharine Walt Disney version.)

Kipling was partly inspired to write the *Jungle Books* by reports, from English commissioners and station agents throughout India, about young village children who had supposedly been adopted by wolf packs. The children were later discovered and rescued, sometimes when they were in their teens. Some of the tales are extremely detailed and accompanied by convincing-sounding supporting statements from witnesses. It is impossible to know how many have a basis in fact, but they are so numerous that at the very least they show there were repeated encounters between the wolves and the village people of India over a long period of time.

What was the world like in which these encounters took place? Alas, not much is left of it, even of the remains of wild India that Kipling knew. But I glimpsed a little of that lost world recently in Ranthambhore National Park in the state of Rajasthan. It was dusk, and I saw a jackal in a clearing, illuminated by a shaft of golden light. Other jackals moved in the shadows nearby. Then, suddenly, the forest was filled with chital deer fleeing from the jackals. Their dappled bodies flashed past only three meters from me. As quickly as they had appeared, they were gone. The jackals, frustrated in their attempt to stalk the deer, melted away to regroup and try again.

For that brief moment I had been taken back to a time when much of India was covered in forests. The woods of that time were alive with the calls of wild peacocks and the roars of hunting tigers. Packs of wolves and wild dogs roamed freely for miles. Leopards, cheetahs and jackals were able to hunt the chital deer without depleting them, while tigers and lions stalked the larger and meatier sambar deer and nilgai antelope. Black sloth bears with striking white markings on their chests prowled in the forests' depths. Pythons and cobras slithered through the undergrowth, and families of magnificent black bucks browsed in the clearings.

This was the magical world that Kipling wrote about. But even during Kipling's time that world was vanishing. Now, especially after the widespread deforestation that followed India's independence, Ranthambhore and the other national parks represent less than one percent of India's original forested ecosystems. And even this remaining one percent is threatened.

Later, in Bandhavgarh Park in central India's Madhya Pradesh State, I was lucky to spot an Indian wolf, one of possibly 2,000 that remain on the entire vast subcontinent. This descendant of Kipling's Raksha, Akela, and Father Wolf was hidden in a dense thicket. She was the first wolf to be seen in the park for over a month. Slim and rangy, she had a keen face and a greyer and much lighter coat than her Tibetan and Asiatic wolf relatives far to the north. Like the Mongolian wolf cub that I was to encounter three years later, she was determinedly gnawing a bone, in this case from an old tiger sambar kill. As soon as she caught wind of us she melted into the undergrowth.

So, if wolves played such an intimate role in Indian life, and interactions were so common, why were dogs apparently never domesticated from Indian wolves?

Mitochondrial DNA evidence shows that Indian wolves form a distinct evolutionary branch from that of the Tibetan and Asiatic wolves to the north. The Indian wolf parted company from the wolves north of the Himalayas almost half a million years ago, and it is distinct enough to merit a different species name, *Canis indica*.

Thus, if dogs had been domesticated from Indian wolves, they would carry a clear genetic signal of their Indian origin. The most likely candidates

Figure 89 An Indian wolf, *Canis indica*, at Ranthambhore. Although they are separated from Asiatic wolves by half a million years of divergent evolution, these wolves would seem to be eminently tamable. But there is no evidence that they were ever domesticated.

for such a domestication would be the half-wild pariah dogs or pye dogs that swarm India's cities and countryside. But pye dogs are not Indian in origin. They are merely another of the many dog lineages that have descended from East Asian wolves.

It is possible that Indian wolves are simply more difficult to tame than Asiatic wolves. But there is another intriguing possibility. Perhaps lineages of dogs were indeed domesticated from Indian wolves, but were then supplanted by more useful breeds arriving from western Asia, just as the Maori dogs of New Zealand seem to have been replaced by European breeds. It will be worth examining DNA from dog bones found in the Indian archeological record to see whether Indian wolves might have been domesticated in the past.

Such a scenario is not entirely far-fetched. Breeds of dog, like human tribes, can go extinct. As we saw, Maori dogs disappeared shortly after the arrival of Europeans, though they may have left some of their genes behind. And the dogs that accompanied the first humans from Asia to the Americas

Figure 90 The behavior of this New Zealand sheep dog will soon be understood at the genetic level.

carried a unique mitochondrial DNA lineage that has since been lost. This lineage disappeared even though it was common in the DNA that has been extracted from pre-Columbian dog remains. Jennifer Leonard of UCLA, who discovered this lineage, suggests that perhaps these native dogs did not have properties that the European colonists found useful.

My guess is that the human–dog bond is so logical, and so advantageous to both sides, that it may easily have been forged more than once, perhaps in India and Europe as well as East Asia. Subsequent loss of these breeds may not seem so far-fetched when we consider how many domestic breeds of other animals have been lost or nearly lost. We may only be starting to learn about the complex history of dog domestication.

Revealing the dog genome

Artificial selection has produced an explosion of different dog breeds. Currently there are more than 400 recognized breeds, ranging from the 1.5-kilogram Chihuahua to the 100-kilogram Great Dane. A visitor from Alpha Centauri would be astonished to learn that such obviously different animals belong to the same species, and that they are (mechanical difficulties aside) fully interfertile.

As of this writing a complete DNA genome from the boxer breed has been sequenced. Less complete sequences have been obtained for 24 other breeds. We know from historical records that most current dog breeds originated only a few centuries ago, and that they tend to be highly inbred. Despite the inbreeding, the DNA evidence shows that the dogs of each breed still preserve a great deal of the genetic variation that they inherited from their wolf ancestors.

Both the maternal and the paternal chromosome sets of the boxer dog have been sequenced. The two sets are separated by three-quarters of a million single-base differences (out of about three billion), showing that there is a substantial amount of genetic variation that still remains within the boxer gene pool. Boxers and poodles differ by one and a half million single-base changes, and preliminary data from other breeds suggest comparable amounts of genetic separation. Most of these breed-specific differences have nothing to do with the artificial selection that generated the breeds. They are simply genetic accidents, the inevitable result of dividing a gene pool up into small subpools. But scientists are beginning to track down some of the more meaningful genetic differences among breeds that dog breeders have produced through their deliberate artificial selection.

For example, it made sense for the first agriculturalists to select for small dogs. Small dogs can sound the alarm against intruders and catch rats as effectively as large ones, and they don't eat as much. A genetic consequence of that selection has now been found by Nathan Sutter of the National Institute of Human Genome Research and his colleagues.[11] Small dogs carry a unique allelic form of an insulin-dependent growth factor gene. The "small dog" allele has been subject to stronger recent selection than the other alleles at this genetic locus that are carried by large dogs. Over the past few thousand years the small dog allele has increased in numbers in dog populations, driven by artificial selection.

Alleles for smallness must have arisen many times in wolf populations by mutation, but they were soon lost because small wolves would have been at a disadvantage. This seems to have remained true even in dog populations up to the time of the agricultural revolution. Before that time, dogs were as large as wolves. Small dogs first appear in the fossil record in the Middle East and Europe about 10,000 to 12,000 years ago, at the same time as the spread of agriculture.

Around that time dog breeders selected for small dogs, which pulled the small dog allele up in frequency. As the allele became commoner it dragged along other genetic differences that lie close to it on the same chromosome. It was this cluster of accompanying genetic differences that provided Sutter with evidence that the small dog allele has increased in dog populations as a result of recent artificial selection.

Enormous behavioral differences among dog breeds have also been produced by millennia of deliberate selection. These differences provide us with great opportunities for understanding the genetics of behavior. Herding, pointing, and retrieving behaviors can now be subject to genetic analysis.

Genes that influence these behaviors will be relatively easy to locate on the chromosomes of different dog breeds.

As a genetic bonus, human medicine is likely to be greatly advanced by the study of dogs. Over 300 dog diseases with a genetic component have human equivalents. Some dogs also exhibit behavioral abnormalities reminiscent of human aberrant behaviors—my favorite is "Springer Spaniel Rage Syndrome." Is this a canine genetic equivalent to human road rage, and do easily enraged humans and dogs carry equivalent genes?

Dogs are likely to point the way to understanding genetic influences on human behavior. They will continue to be man's best friend, even at the genetic level.

Figure 91　Mongolia's magnificent Gorkhi Valley provides summer grazing for yaks, horses, and goats.

Yaks, zebus, and aurochs

At the moment the preponderance of evidence is that present-day dog breeds eventually emerged from a series of dog domestications that took place in Asia about 15,000 years ago—though it is possible that domestication began earlier. This scenario suggests a world of nomadic communities made up of children who kept bringing home wolf puppies and patient adults who tolerated the inevitable snarls and bites. Other centrally important domestications that have shaped the modern world took place for strikingly different reasons. One of these domestications involved the yak.

Terelj is a huge area of mountains, valleys, and grasslands that stretches to the northeast of Ulaan Baatar, the capital (and only major city) of Mongolia. When I was there in July 2008 the countryside was ravishingly beautiful, a blazing green sprinkled with Edelweiss and other wildflowers. The Gorkhi Valley in the heart of Terelj is dotted with weathered sandstone pinnacles, left behind by the great river that flowed there in a wetter past.

Nomad families come to the area for the summer grazing. Their herds of horses and yaks provide milk, and the goats produce valuable cashmere wool.

The long-haired yaks are especially endearing animals. Calm and phlegmatic, they stick out their blue tongues thoughtfully as they move across the landscape like large hairy hassocks.

Wild yaks that live on the high Tibetan plateau to the south have long coarse coats that come in basic black. For a variety of physiological reasons wild yaks do not thrive at lower altitudes such as the Gorkhi Valley.[12]

Wild yaks have special circulatory adaptations that, during pregnancy, increase blood flow to their placentas and ensure that the baby yaks receives plenty of oxygen during their critical early development. The human inhabitants of Tibet, who have had at least 10,000 years to adapt to their high-altitude plateau, have evolved a yak-like circulatory mechanism to help nurture their babies in the womb. Native Tibetans' babies are heavier at birth and more likely to survive than those of Chinese women who have come to settle on the Tibetan plateau.[13]

Figure 92 At Gorkhi, purebred yaks, *Bos grunniens*, have been crossbred with cattle. The hybrid animals are well adapted to the intermediate altitudes of central Mongolia.

Because wild yaks are adapted to high altitude, the animals that have done well in these rich grasslands tend to be hybrids. The yaks in the Gorkhi Valley have been crossbred with cattle, and have blossomed into a variety of colors. The male hybrids are sterile, but the females are fertile and can be crossed further. The hybrids are not as thickly pelted or as physiologically well adapted to altitude as their Tibetan relatives.

Peter Savolainen and Chinese collaborators have carried out genetic studies of yaks.[14] Their mitochondrial DNA surveys of domesticated and wild yak populations show that the domesticated yaks have inherited two different lineages of wild yak mitochondrial DNA. These workers have followed the approach that Savolainen had earlier used in his dog study to locate the most likely site of dog domestication. They scanned yak populations from all around the Tibetan Plateau and found that the most diverse populations were on the border of northeast Tibet and China's Qinghai Province, 1,500 kilometers south of the Gorkhi Valley. This area, they concluded, was the most likely site of yak domestication.

The domestication of yaks seems to have been much less complicated than the domestication of dogs. Both the genetic and the archeological evidence point to a single domestication event that happened approximately 10,000 years ago. At around that time the grip of the most recent Ice Age had begun to ease. As glaciers started to retreat, the higher valleys of the great Tibetan plateau emerged from their long deep freeze. The tough nomad tribes who ventured into the mountains to explore these new areas encountered yaks already thriving there. Unsung geniuses among them realized the potential of yaks for milk and meat, and found ways to confine and herd the animals.

Dogs must certainly have played a role in the domestication of these valuable yak herds. They helped to track down lost animals and they protected the herds from wolves and, in the more forested regions, Siberian and Caspian tigers.

Much later, domestic cattle from India and the Middle East were introduced into the area, where they hybridized with the yaks. It soon became clear that these hybrid animals could do well at altitudes where both yaks and cattle felt uncomfortable.

The most genetically diverse domestic cattle breeds currently live in the Middle East and northwest India, marking their probable points of origin. Domestic cattle had a more complex origin than domestic yaks, one that was bound up with religious beliefs as well as mere utility.

The Middle East of 10,000 years ago was a transition zone between dense European forests to the north and west and an African-like savanna to the south. Because these ecosystems were both still largely intact, the zone provided a dramatic collision between these two biological worlds, even more dramatic than the ecological collision zone that we saw in Chapter 2 when we visited the present-day Evolution Canyon of northern Israel.

In the forested parts of this interface lurked the terrifying aurochs, the wild forerunner of domestic cattle.[15]

The aurochs itself is now extinct, but it used to be one of the most dangerous animals inhabiting the dense forest that covered most of Europe. Aurochs grazed in small open areas within the forest, but in their search for more grass they continually moved from one clearing to the next. As a result the people who lived in the forest never knew when or where they would stumble on these dangerous animals.

We know, from DNA taken from aurochs bones, that these progenitors of cattle diverged from other bovid lineages at least a million years ago. They were genetically diverse and scattered over a wide geographic range, from Britain to the Middle East and North Africa. But as their forests disappeared and human hunting pressure increased, their numbers plummeted. The last aurochs died in captivity in Poland in 1627. Its skull, adorned with magnificent lyre-shaped horns, has been preserved.

When Caesar battled German tribes in the dense forests at the fringes of Rome's empire he encountered many aurochs and was impressed with their ferocity. He remarked that they seemed impossible to tame even when they were young.[16]

Despite Caesar's observations, some aurochs had in fact become part of human lives millennia earlier. In Anatolia, their skulls and horns have been found buried in earthen walls dating to 12,000 years ago. Aurochs appear in Neolithic cave paintings, and they played a role in Mithraism and other Mediterranean religions. An aurochs is likely to have inspired the story of the Minotaur.

The fierce aurochs were initially tamed by the herdsmen and proto-farmers of western Turkey and northern India, who were faced with a daunting task. Mature aurochs weighed twice as much as the average domestic cattle of the present time. It seems likely that, as in dog domestication, taming began with the capture of young animals. The early herders would have selected the animals that matured into the smallest and mildest mannered adults, and let the dangerous ones go.

Two types of modern cattle are derived from the aurochs, the familiar *Bos taurus* that is adapted to wet climates, and an Indian breed called the zebu that is superbly adapted to near-desert conditions.

Zebu were probably first tamed in or near the Thar Desert in Rajastan in northern India. They are rare in countries with plentiful rainfall, but have great value in deserts and semi-arid regions. They are the commonest type of cattle in Asian deserts, in southern Africa, and on the drought-plagued island of Madagascar.

And, as I can attest, the zebu cattle have retained the great curving lyre-shaped horns of the aurochs. On a busy street inside the walled fort of Jaisalmer, near the Pakistan border, I was tossed to the side by an irritated zebu that claimed the right-of-way. I did not reflect on the evolutionary implications of our encounter at the time—I was too busy grabbing the zebu's horns to keep from being skewered. But afterwards I realized that I was grabbing a piece of evolutionary history. I was being roughed up by this zebu not far from the spot, 10,000 or more years ago, where the zebu had first been domesticated.

I encountered zebu again a year later, in the Ambalavao market in central Madagascar. Cattle herders have brought their zebu to this thriving market town for centuries.

Zebu cattle are essential to the way of life in many parts of the underdeveloped world. The Malagasy zebu are close relatives of the Indian zebu and show little sign of having been interbred with other strains of domestic cattle. Thus they may not have originated in nearby southern Africa, where there has been much interbreeding of zebu with other strains of cattle. They may instead have arrived centuries ago by a now-lost trade route from Asia.

Regardless of how they arrived, the impact of the zebu on Madagascar's fragile ecology has been devastating.[17] This is because they are able to survive on some of the most abused land on the planet, the sparse grasslands of Madagascar's central plateau that a thousand years ago were covered with dense forest.

The people of this vast plain are desperately poor. Each year the herdsmen burn huge areas of the plateau in order to coax forth a scattering of fresh grass.

When I walked over burned areas of the plateau, with their crisped, utterly exhausted soils, it was like crunching across the surface of a gigantic crouton. The seared landscape is rapidly eroding, washed away by riv-

Figure 93 Zebu cattle with their elegant lyre-shaped horns are brought by their herders to this market in Ambalavao, central Madagascar. The zebu thrive under conditions where most domestic cattle breeds do poorly.

ers that oscillate between bone-dry and raging torrents. The rivers carry the red soil into the Indian Ocean and Mozambique Channel. When astronauts look down at Madagascar from orbit they describe an island that seems to be bleeding to death.

Madagascar's exploding human population is primarily to blame for the island's slow-motion ecological disaster. It is ironic that the artificial selection that produced the superbly adapted zebu gave them the ability to survive on the sparse grasses that emerge after the massive burns, but has also contributed to this devastation.

The domestication of zebu is not the only domestication that has had unintended consequences. At around the same time as the aurochs was being

Figure 94 Madagascar's central plain has been repeatedly burned to provide fresh grazing for zebu, making it one of the most abused regions of the planet.

tamed, goats and sheep were being domesticated from wild populations in central Asia, and perhaps the Fertile Crescent of the Middle East.[18]

Of the two, goats have done the most damage, because they pull up entire plants and eat their roots. Overgrazing by goats has contributed to the devastation of ecosystems in the Mediterranean, the Middle East, and Africa.

The origin of goats is a little unclear. Several wild species contributed to their ancestry. Cashmere goats, which are adding to the prosperity of Mongolian nomads at the same time as they severely overgraze the grasslands of Inner Mongolia, have been domesticated from a distinct lineage of wild goat. Whether that lineage was actually a different wild goat species, or whether

Figure 95 Tim Short and his family at New Zealand's Mount Tutu Nature Reserve show off their unique Mount Tutu sheep, bred over a period of twenty years from a Romney–Perendale cross.

some ancestral population of wild goats was genetically diverse enough to give rise to all present-day goats, remains to be determined. Many wild goat species have recently gone extinct, and it will be necessary to obtain DNA from their remains in order to untangle the details of how goats were domesticated.

Sheep present a similar picture, though here the point of origin is more definitely the Fertile Crescent. Again, multiple independent domestications are likely, and more than one wild sheep species may have been involved.

Figure 96   The small group of Romney–Perendale that Tim started with in 1988 to produce his astonishing breed. (Courtesy Tim Short.)

Domestic sheep are astoundingly diverse morphologically. It is possible in just a few generations to produce a new sheep variety, as I discovered when I visited Tim Short's eco-reserve on New Zealand's North Island and encountered his spectacular Mount Tutu breed.

The partial taming of the Asian elephant

The aurochs and many of the possible ancestors of goats and sheep are now gone, driven to extinction during the same period that their domesticated descendants began to change the world. The details of their domestication have largely been lost. But sometimes we can still catch a domestication in progress.

In South and East Asia, elephants have repeatedly been tamed to serve as laborers and as fearsome engines of destruction in warfare. The tamed elephants were recruited from the herds of wild elephants that were once plentiful throughout the Asian tropics.

The landlocked country of Laos is at the very center of what used to be the wild elephant distribution. It encompasses the heart of the Lan Xang empire, which thrived from the fourteenth to the eighteenth centuries. The empire's name, which means "million elephants," was bestowed by its founding king Fa Ngum. His goal was to terrify his enemies with the possibility that a million elephants were about to crush them.

There were certainly plenty of elephants in Laos during the empire, though perhaps not a million. There is also no doubt that the numbers of wild elephants have plunged since the time of Fa Ngum. Present-day estimates lie between as many as 2,000 or 3,000 down to as few as 200.

At the same time as wild populations of elephants have declined, the meaning of wild has changed. In the course of numerous attempts at domestication, elephants have repeatedly been released or escaped back into the wild. Many of their progeny have been recaptured, often with the aid of tamed elephants trained to act as lures. As a result, in India, Myanmar, Laos, Cambodia, and Thailand, elephants have become a mixture of feral and domesticated. After thousands of years there are unlikely to be any truly wild elephants left.

The English and Australian researchers C. M. Ann Baker and Clyde Manwell have suggested that the result of all these attempts at taming has been a kind of partial domestication.[19] The effects of taming on elephants brought in from the wild can immediately be reversed when the penned elephants escape. The result is an equilibrium between wild and tamed behaviors, like a chemical reaction that fails to go to completion. But partial though these repeated domestications have been, they must have had an impact on the Asian elephant gene pool. The elephants that could be tamed most successfully survived and had offspring, while those that were intractable were driven to the furthest forests or killed.

Despite millennia of association with humans, and despite repeated attempts at selection for tameness, Asian elephants remain dauntingly dangerous. In Bandipur National Park in southern India I saw a huge tusker striding down the side of the road. He had been giving himself a dust bath, and his back was still covered with reddish earth. He looked neither to the left nor to the right as he moved swiftly along. Six weeks earlier he had killed an incautious tourist from Bangalore, who had stepped out of his car to take his picture. I snapped the elephant's picture too, of course, but without leaving my vehicle.

Figure 97 A rainbow arches over the temple-filled city of Luang Prabang. The lush hill country of Laos, at the center of Asia's elephant population, was once home to immense numbers of wild elephants.

Such uncontrolled behavior is not the elephant's fault. Male elephants periodically undergo *musth*, a time when testosterone peaks and glands on the sides of the head become activated and ooze foul secretions into the elephant's mouth. As if this were not enough, it appears that the swollen glands may cause blinding headaches. Elephants in *musth* have to be confined and in some cases chained down.

The unpredictability of elephant behavior means that the wild and domestic gene pools of the Asian elephant have remained commingled, and that it is impossible to determine when elephant domestication began. It seems likely, however, that attempts at domestication started thousands of years ago.

Untamable African elephants?

The story of elephants in Africa is puzzlingly different. We have no recorded instances of any of the native peoples of sub-Saharan Africa trying to tame African savannah or forest elephants.

The elephants that lived north of the Sahara may have been more tractable. Hannibal, the North African general who almost succeeded in conquering Rome two centuries before the start of the Common Era, took more than thirty elephants with him in his march across the Alps. These elephants may have been a local North African race, now extinct. Or they may have been from south of the Sahara—a Carthaginian coin from that time shows an unmistakably big-eared African-type elephant. But Gavin de Beer has pointed out that Carthage is known to have traded with Egypt, and that the Egyptians had captured Indian elephants during their earlier wars in Syria. So there is a good possibility that Hannibal's fighting elephants were actually the more tractable Indian species.[20]

Figure 98 A killer elephant in Bandipur National Park in southern India. Elephants in India have a mix of wild and tame ancestors, and after thousands of years of attempted taming by humans none are still truly wild.

Figure 99 This domesticated elephant, at an "elephant temple" near Cochin on India's southwest coast, is going through a period of musth and has to be restrained. Such unpredictable behavior has made it difficult to complete the domestication of Indian elephants.

Humans have hunted elephants in sub-Saharan Africa for a long time, though they did not drive them to extinction. Elephants in more temperate latitudes were more vulnerable, however. Todd Surovell of the University of Wyoming and his colleagues have used the distribution of hunting and butchering sites to show that, as humans spread north into Eurasia and finally into North America, they probably did hunt the local elephants to extinction.[21] Butchering sites formed a thin line that moved north over time, showing that humans killed as they went, wiping out the elephants quickly and with unnerving efficiency. Ten thousand years ago, not long after humans arrived in North America, four genera of elephants and mastodons that had been thriving there disappeared.

In contrast to the slaughter in more temperate latitudes, humans and elephants coexisted in Africa. Why were the elephants not tamed? Are sub-Saharan elephants really too savage, or did it simply not occur to anybody to try? There is no doubt that domestication of African elephants requires a great deal of patience, but it seems that it is not impossible.

Recently, in the western part of Botswana's Okavango Delta, I watched as a huge female African elephant plunged from the shore of one of the islands and swam swiftly across a wide branch of the river. She held her trunk up like a snorkel as she swam. Elephants are excellent swimmers, and can easily travel across the delta even when it is in full flood.

This elephant, one of many living in the delta, did not look as if she would brook any sort of interruption of her natatorial activities. Nonetheless, guides at the Abu Camp, a tourist facility a few kilometers to the west of where I saw this elephant swimming, have managed to train several female

Figure 100 An African elephant swims easily from one island to another in Botswana's Okavango Delta. African elephants have recently been tamed, which makes it more puzzling that they were not tamed earlier during the long history of humans on the continent.

savanna elephants. They now cheerfully carry tourists across both land and shallow water, allowing their passengers to observe the wildlife of this rich landlocked delta from their gently rocking backs. Unlike that poor tourist in Bandipur who was killed by a feral Indian elephant, none of the tourists at Abu have so far been harmed.

At the Garamba National Park in the Congo there is an African Elephant Domestication Center. Poaching and invasion by Sudanese guerillas have repeatedly disrupted the park's operations, but before the disruptions some elephants were trained to carry tourists. It is possible that the Center will be established again if peace returns to the area—as of late 2005 there were still 1,200 elephants in the southern part of the park despite continued bushmeat and ivory poaching.[22]

Intriguingly, genetic studies have shown that the elephants of the Okavango have interbred with forest elephants, and the trained elephants in the Congo's Garamba were also forest elephants. It is possible that these shy denizens of West Africa's rainforests may have milder temperaments than savanna elephants, which may explain the guides' success in teaching them to carry tourists. But the failure of native tribes to tame African elephants, either forest or savanna, is an unusual gap in what has generally been a series of success stories of domestication around the world.

In contrast to Africa, the inhabitants of Eastern Asia have been strikingly successful in the taming of animals. One of the most remarkable events, one that changed history and even changed our own gene pool, is the domestication of the horse.

Horses, camels, and the compression of space

The taming of horses triggered one of the most transformative events in human prehistory. For the first time people could travel faster and further than it was possible to walk or jog. Horses compressed the world, so that in a single lifetime an adventurous individual or an entire tribe of nomads could travel for thousands of kilometers. Horses were as liberating for our ancestors as cars are to us.

The origins of modern horses are a bit mysterious, because they have only one close living relative. But it is likely that they were first domesticated in eastern central Asia, in the same huge region of open plains and mountain valleys where dogs and yaks were also tamed.

The only living relative of the modern horse is the Przewalski or Mongolian horse of central Asia. The horse was named after Polish naturalist Colonel Nikolai Przewalski, who explored the area in the 1870s and 1880s. He obtained a skull and hide of the horse that bears his name and shipped them to Europe.

These small tough horses, with their dark legs, tan bodies, and wire-brush manes that stand straight up, have survived by a fluke. About a hundred and fifty of them were captured and sent to zoos in Europe at the beginning of the twentieth century. But they continued to be hunted in their homeland. The wild Przewalski's horses were finally driven to extinction in Asia half a century ago. The last of them—and indeed the last truly wild horse anywhere in the world—was glimpsed in 1969, in the barren mountains of western China, not far south of where I saw the wolf cub.

The Przewalski's horses that survived in zoos have been crossed to domestic horses, so that no completely unmixed strains remain. Some mostly wild strains of these horses were recently reintroduced into Mongolia and China. My colleague Oliver Ryder, who was involved with the Mongolian reintroduction, has shown that the purest Przewalski's horses today are descended from only four zoo females.

The history of the Przewalski's horses is intertwined with that of the domestic horse.[23] Mitochondrial DNA studies show that the wild Przewalski's horses and domesticated horses share the branches of a single family tree. Only one unusual mitochondrial genome has been found in a Przewalski's horse. All their other mitochondrial genomes resemble those of domestic horses. It seems that the hybridization between domestic and Przewalski's horses has been extensive, so that some of the Przewalski genetic information may have been lost.

The data are not yet complete enough to see whether domestic horses could have had a single origin. There is, however, one striking piece of evidence suggesting that something dramatic happened during or after horse domestication. Domestic horses have one fewer chromosome in their set of chromosomes than Przewalski's horses (just as we have one fewer chromosome than our closest living relatives the chimpanzees). Either a pair of chromosomes fused together in the domestic horse lineage, or a chromosome split into two in the Przewalski lineage. The resulting chromosomal mismatch does not seem to give problems in crosses between these two groups of horses—the hybrids are healthy and fertile.

There are two possible explanations for how this difference in chromosome number arose. The first is that the real ancestors of domesticated horses were not Przewalski's horses at all, though they were undoubtedly closely related to them. Instead they were a different species, with one fewer chromosome than the Przewalski's horses. This wild horse species, like the Przewalski's horses, also ran free across the steppes of Central Asia.

In this scenario, groups of Asian nomads domesticated some members of this wild species several thousand years ago. Then, as the domestic horses and their nomad owners proliferated, their wild ancestors were driven to extinction, as had happened to many horse species before them. The only Asian wild horses left were Przewalski's horses. They too were being driven to extinction, but unlike the true ancestors of the domesticated horse a few of them luckily survived in the Munich and Prague zoos.

The second and much less likely possibility is that domestic horses are descendants of a mutant strain of Przewalski's horse, one in which two chromosomes had become fused. This scenario requires the close juxtaposition of several unlikely events: a chromosome fusion that spread through a local population of Przewalski's horses, followed by the domestication of some of the animals that carried the fusion, and finally the extinction of all the wild members of the Przewalski's horse group that carried the fused chromosome.

Both of these scenarios are possible, but I am inclined to wield Occam's razor and slice away the second less likely one. If the first scenario is the right one, it demonstrates once again the enormous impact that humans have had on the wild ancestors of our domesticated species. We humans may have driven not one wild Eurasian horse species to extinction, but two!

My guess (and it is only a guess) is that all the wild horses of Europe and Asia were on their way to extinction well before some of the survivors were domesticated. We will see in Chapter 8 that humans have intensively hunted wild horses for hundreds of thousands of years. The remains of large numbers of wild horses from 400,000 years ago have been found at one small site in Germany. They were butchered by close relatives of ours, the pre-Neanderthals. As a result of such intensive hunting, there were probably only a few populations of wild horses left by the time some of them were finally domesticated. And the wild remainder, made up of animals that were less easily tamed, were hunted down across the open steppes where they had no refuge. The last of them were eaten long before naturalists arrived on the scene.

The camels that also play a central role in Mongolian life have undergone a similar saga of domestication accompanied by the extinction of their wild progenitors. The two-humped Bactrian camel is a tough animal found today throughout the cold deserts of central Asia and as far west as Kazakhstan. It seems to have been even more widespread in the past, extending its range all the way to the Indus valley, but in many areas it has been supplanted by the one-humped dromedary. Its closest wild relatives now consist of a few thousand wild two-humped camels roaming the Mongolian–Chinese border regions. The wild camels have smaller humps and longer, thinner legs than the domestic variety.

The two-humped Bactrian camels carried silks, spices, gold, and jewels along the Central Asian Silk Road, along with the traders who dealt in them. As a matter of personal taste, I much prefer riding two-humped camels. Nestling comfortably between their humps is like riding a La-Z-Boy recliner with legs (and diarrhea). It is an entirely different experience from balancing on the top of a dromedary's single hump as it snarls, spits, and lurches across the desert through a series of increasingly uncomfortable forward speeds.

A group of Chinese investigators from Beijing, Shanghai, and Inner Mongolia have now shown that, like the horse, the Bactrian camel was most likely to have been domesticated in the region of Mongolia and western China that lay at the heartland of the later Mongol empire. They also showed that

Figure 101 Horses and their riders gallop towards the finish line of a ten-kilometer race across the vast gravel desert in western Mongolia.

domestic two-humped camels from throughout their range are all closely related to each other. They are surprisingly distinct from their wild relatives. Judging by the genetic differences between them, the domestic and wild camels are at the very least different subspecies.[24]

Like horses, domestic camels must have originated from a species or subspecies that has now gone extinct in the wild. The only wild Bactrian camels that still survive are that parallel group of wild two-humped camels that (like Przewalsky's horse) were not domesticated. The remaining small herds of these wild camels tend to flee (with good reason) at the first sign of approaching humans.

Horses and camels are central to life in Mongolia today, and dominate much of its history. But it is striking that such an obvious opportunity for domestication was not seized on by early peoples in other parts of the world. In Europe, the pre-Neanderthals, Neanderthals, and later early modern inhabitants simply slaughtered the horses that they found there, though the early modern humans did at least draw pictures of them on cave walls. And the first peoples who migrated into North and South America 13,000 years ago are likely to have driven the native American horses and camels to extinction, just as they did the native American mastodons.

In contrast, when horses were reintroduced to the Americas by the Conquistadors in the early 1500s, their fate was quite different. They were spread to the north from Mexico by the Europeans and some Indian tribes. Some were eventually stolen and used for hunting and warfare by the plains Indians. Within a generation or two, the Comanches and other tribes had seamlessly merged horses and firearms into a superb cavalry that often won pitched battles against the Europeans.

Why did the first Asian migrants to the Americas, coming directly from a culture that had already tamed wolves, not tame the horses they found on the new continents? We know nothing about the behavior of *Equus scotti*, the native American horses that the early migrants from Asia found living on the American plains. Perhaps they were untamable, like zebras.[25] But we do know that the horses brought by the Europeans to the Americas were the result of thousands of years of selection. They could be approached more easily than truly wild animals such as zebras. Their appearance and behaviors may have called out to the plains Indians: "Hey! I'll bet you too can ride me!"

Horses and the human gene pool

The day before the eclipse, we took part in a festival that the people of the nearby town of Khovd had put together to celebrate the event. Wrestlers strained in mock combat and a group of elegantly costumed women danced to the music of a Chinese lute, while a flock of steppe buzzards dipped and swooped above the enthusiastic crowd.

Afterwards everybody surged to the edge of the fairground to watch the finish of the great horse race. I found a vantage on a little escarpment.

The grass of the fairground petered out into an enormous gravel plain that stretched out to blue mountains on the horizon. At first nothing happened. Then, at least three kilometers away, a tiny cloud of dust appeared. As we strained to see through the heat haze the racers themselves gradually became visible, still embedded in the cloud that they had generated. The excitement of the crowd mounted as individual horsemen became recognizable.

The winners raced toward us and fanned out as they reached the ill-defined finish line. And it became clear that these horsemen were not grizzled veterans of the steppes. They were young boys of ten or twelve, who had ridden their tough little horses bareback at top speed across ten kilometers of parched country. These tough-as-nails kids and their ponies, seamless super-organisms of horse and rider, rode by in an impromptu victory parade as the onlookers cheered.

The use of horses has been central to Mongolian identity for millennia. And now there is evidence that the possession of these tough, tiny horses helped to give some of these people of central Asia an enormous evolutionary advantage. The result was a pronounced change in the gene pool of the people who lived in the entire region.

An individual's fitness in the evolutionary sense is defined by how many children he or she has. And it seems that some Mongolians were exceedingly fit. These individuals, in part because of their horses, had an amazing number of children.

The evolutionary success story of these extremely fit nomads can be traced through the Y chromosome, the chromosome that is passed from father to son and that determines whether an offspring will be male. Like the mitochondrial chromosome, part of the Y chromosome does not recombine with other chromosomes. It simply accumulates mutations over time. Family trees can be built using this part of the Y in exactly the same way that family trees can be built using mitochondrial chromosomes.

Approximately a third of the males in Mongolia have a particular type of Y chromosome. We can infer from this large number that many of their recent male ancestors, who also carried this Y, must have been extraordinarily fecund.[26] The rest of the Y chromosomes in the Mongolian population, and indeed most Y chromosomes in Asia and elsewhere, come in a wide variety of different types, none of which is especially common. The males who carry these other Y chromosomes have inherited them from less fecund recent ancestors.

Genghis Khan, the most successful nomad of them all, rode his way to domination over half the known world on the back of a sturdy Mongolian horse. He was assisted by two important inventions that had been made earlier, the stirrup and the short recurved bow that allow a horseman to fire arrows in any direction while riding. It is strongly suspected that he and his immediate male relatives carried the Y chromosome that is now so common in Mongolia and the regions nearby.

Figure 102 The young winners of this bareback race parade at the finish line.

Additional circumstantial evidence reinforces the likelihood that this common type of Y chromosome was carried by Genghis Khan, or at least by his close relatives. The Hazara, who live on the Afghan–Pakistan border far to the west of Mongolia, have a proud tradition that their tribe was founded by Genghis Khan. Their tradition may be based in fact. Half of the Hazara males show this unique Mongolian genetic pattern on their Y chromosomes, and none of the other groups of peoples who currently live near the Hazara carry this pattern.

Like the mitochondrial Eve, this unique Y chromosome can be traced back to a single individual. The current best estimate is that the chromosome probably arose by mutation from another Mongol Y chromosome about a thousand years ago, two centuries before Genghis Khan. The odds are good that Genghis Khan was one of the descendants of this "Mongolian Adam," because there are no historical records of any other individuals living at that time and place who had access to as many women as Genghis Khan and his immediate descendants. He was not the originator of this Y, but he spread it like crazy.

Although Genghis Khan had only one empress, the redoubtable Börte, he had dozens of other wives. He also raped or made love to hundreds of captive women who were brought to him from the far-flung outposts of his empire, which at his death stretched from western Asia to northern China. His grandson Kublai Khan, who conquered the rest of China and established the eastern Mongol empire, had a harem of thousands of women. He received shipments of thirty virgins a month from all over the empire.

A direct proof of whether Genghis Khan actually carried this Y chromosome awaits the discovery of his tomb, which is hidden somewhere in the mountains of northeastern Mongolia. A small piece of one of his bones should yield enough DNA to settle the question.

The tomb will be hard to find. A lurid legend told by people still living in the area recounts how the tomb's site was kept secret to protect its supposed vast treasure from grave robbers. The soldiers who accompanied the Khan's remains killed anyone they encountered en route. Once the burial was completed, they then killed all of his funeral attendants and trampled the site with horses. They even diverted a river to flood the site. Unknown to these soldiers, and to make assurance doubly sure, Genghis Khan left a posthumous order to have the soldiers themselves killed on their return home, so that the secret of the tomb's location died with them. You are free to believe all this or not, as you wish, but the legend has inspired generations of explorers to search for the tomb.

Albert Lin at my home institution has explored the most likely area that conceals Genghis Khan's tomb by satellite photographs and on horseback. He has found traces of hundreds of tombs and other structures that postdate Genghis Khan. Fallen trees have torn up the ground in many places and revealed roof tiles of substantial buildings, showing a Chinese presence in the area subsequent to the time of Genghis Khan. Much archeological detective work will be needed to disentangle the complex history of the region.

Approximately a thousand years ago, perhaps as little as two hundred years before Genghis Khan, an unknown Mongolian Adam had the only copy of this Y chromosome. It is a fascinating exercise to work out how reproductively successful he and his descendants must have been.

At the time of that unknown Mongolian the world's population was about one-twentieth as large as it is today. This means that on average each person who lived then has twenty descendants living today, half of whom are male. (Of course, the luckier of these ancestors have more surviving descendants than that, and many unlucky ones left no present-day descendants at all.) But when we count up the numbers of those unusual Y chromosomes carried by males at the present time, we find that this unknown Mongolian had, not ten, but 16,000,000 male descendants! This chromosome is truly in a class by itself.

At first glance, you might think that the unknown Mongolian was one and a half million (16,000,000 divided by 10) times as successful at reproducing as the average male of that time. But the advantage was not really that immense. The carriers of the unknown Mongolian's chromosome did benefit from an increase in their reproductive success compared with the rest of the population, but the difference was compounded over a thousand years. Like the multiplying advantage of compound interest, even a relatively small increase in fitness each generation can produce dramatic results over so much time.

Consider ordinary males living a thousand years ago who were not lucky enough to carry the unknown Mongolian's Y chromosome. On average they increased in number by a piddling 4.5% per generation (assuming twenty years per generation). This small reproductive rate of return was enough, through the magic of compound interest, to yield an average of ten males at the present time for every male who lived then.

The lucky unknown Mongolian, however, benefited from a much higher rate of interest. His male descendants increased on average by 33.2% per generation, or 1.65% per year. This would not have been a rate high enough to excite the titans of Wall Street. But it would have been enough over a thousand years to yield 16,000,000 male descendants.

The huge increase in the unknown Mongolian's descendants did not take place at a uniform pace. It seems likely that Genghis Khan's penchant for rape and pillage, and Kublai Khan's nightly deflowering of one or more virgins, gave the unknown Mongolian's Y a big boost near the start of its career. But it also turns out that the reproductive efforts of the Khans, impressive though they were, could not have been enough to account for the astounding number of their descendants at the present time.

Let us suppose that by the time Kublai Khan died, a little more than 700 years ago, his efforts and those of his lubricious relatives had increased the number of the unknown Mongolian's male descendants to 10,000. This would be a gigantic increase in numbers from the time of that solitary Mongolian, especially if we assume the most likely estimate that he had lived only 200 years before Genghis Khan. Then, if the reproductive success of those descendants had subsequently fallen back to the worldwide average of 4.5% per generation, by the present time they would number only 48,000. It is clear that many Mongolians in addition to the Khans themselves benefited reproductively from their conquest of much of the known world.

Genes, environment, and domestication

As the moon completely masked the sun late on the afternoon of August 1 and the sun's pearly corona suddenly flashed into view, a gasp went up from

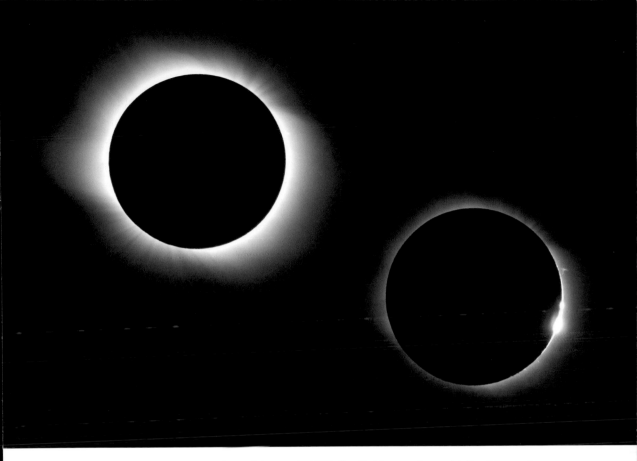

Figure 103 The total eclipse of August 1, 2008. The first image shows details of the sun's corona at the height of the eclipse. I took the second at a faster exposure so that much less of the corona is visible. In the second picture the sun is just starting to shine through a valley of the moon, and you can see the red lower solar atmosphere. The solar eruptions called prominences are few, small, and pale in color, because the sun was in a quiet phase at the time of the eclipse.

the assembled spectators. The Bactrian camel that I had hired for the occasion was unimpressed, but both the Mongolians and the foreign visitors to this remote corner of their country stared at the sky in fascination until the sun re-emerged, shining through the valleys on the rim of the moon to give a sparkling diamond ring effect.

We had all been able to gather at this remote place because of a confluence of science and technology. We had benefited both from a precise prediction of the eclipse's path and from the jet planes that had carried us here. In the space of a single remarkable day we had spanned much of human culture. We had seen how dogs might have been domesticated by nomads on the steppes 15,000 years ago, And we had watched the eclipse unafraid, benefiting from the scientific approach to the world that has made our lives richer by banishing superstitious notions about monsters eating the sun.

The domestication of animals, starting with the East Asian domestication of the dog, helped to provide us with the resources to build our present-day technological civilization. How have we ourselves changed, driven in part by the results of these domestications, and how have we changed the world?

The genetic changes to our own species have been numerous during this time, but the influence of animal domestications on these changes is complicated to say the least.

We have seen how the unusual reproductive success of the unknown Mongolian's descendants continued even beyond the time of Genghis and Kublai Khan. Many of their descendants also enjoyed a higher than average evolutionary fitness. Some of the boys that I saw racing their horses across the valley floor must have been carriers of the unknown Mongolian's chromosome. They had benefited from a millennium of reproductive success that has been unequaled in modern human history.

Is there some gene, on the Y chromosome or elsewhere, that is responsible for this high fitness? Almost certainly not. There are few genes on the Y chromosome to begin with, and the genes on our other chromosomes are mixed and scrambled each generation. Any association of a favorable gene on the other chromosomes with the unknown Mongolian's Y chromosome would soon have been broken. But the men who carried that Y chromosome had access to, or proximity to, power and privilege for many generations. That power was initially generated by the exploits of Genghis Khan and his horsemen.

The taming of the dog, the yak, the Bactrian camel, and probably the horse in this remote and challenging region of the world set the evolutionary stage for one of the most cataclysmic events in human history, the sweep across the known world of the Mongol horde. There is no doubt that the accumulated effects of such social and cultural changes over the millennia have had an impact on our gene pool.

New molecular evidence shows that the rate of our evolution has accelerated over the past 80,000 years.[27] This wave of genetic change started in Africa and swept through Europe and Asia. Some of these changes, such as the ability to tolerate the milk sugar lactose in the diet, have been aided by the consequences of animal domestications, which have influenced many aspects of how we live.

But, as we have seen over and over again in these chapters, at the same time as our own lives have changed and become richer, we have dangerously degraded our environment.

Just in the area of animal domestication alone, our ancestors have driven to extinction the closest wild relatives of cattle, horses, goats, and camels. Much of the store of genetic variation that might have contributed to the future evolution of the species on which we depend has been lost.

And we have flooded fragile ecosystems with unsustainably large numbers of our domesticated animals. As a result, deserts are invading grasslands in Africa and central Asia. In the lands that surround the Mediterranean Sea entire ecosystems have been "goated" into oblivion. Because of overgrazing and unsustainable agricultural methods, the people living on Madagascar's central plateau can only eke out a precarious existence.

Despite all these ecological disasters, I cannot help but wonder whether our long association with domestic animals may have had some positive effects. The process of domestication may have put a premium on behaviors such as patience and empathy. Perhaps someday soon, as we continue to analyze how our genes have changed, we will find out which of our own recent evolutionary changes have been driven by our intimate association with the many other animals on our planet that we have tamed and even befriended. Has there been, as a result, an evolutionary increase in kindness? I hope so.

7

The Great Migration

Figure 104 We flew in the personalized airplane of Australian bush pilot Bob Bates to the Bensbach River basin in southern Papua New Guinea. This lush area is filled with the same animals and plants that are found in Australia to the south, but it benefits from a year-round water flow from New Guinea's mountains.

When you float down the Bensbach River in southern New Guinea, you might think that you are in Australia. But in fact the tree- and grass-covered plain nurtured by the Bensbach resembles the Australia of some wetter past, not the rust-red and often forbidding Australia of the present.

The flat coastal plain of southern New Guinea is watered by the flow of year-round rivers such as the Fly and the Bensbach that originate in the island's mighty Central Range. These rivers nourish a countryside far lusher than the flat Australian plains across the Torres Strait to the south. The coastal plains of Australia, equally pancake-flat, have no mountains nearby to feed their more seasonal rivers. When sea levels rose after the last Ice Age, they cut northern Australia off from the rivers of New Guinea's central massif.

The Bensbach winds through green beds of reeds, where Australian pelicans swim like graceful swans in an English park. Spangled kookaburras laugh from the trees, and clusters of brown night herons watch for fish from the upper branches. Collared kingfishers flit along the river's banks. A fish eagle scoops a huge fish from the river and flies away. And, to complete the picture, mobs of wallabies hop into the lush undergrowth after drinking from the river.

The scenery along the river must have been very like the scenes that greeted the first human migrants between New Guinea and Australia—with one notable exception. Then it was possible for people migrating south to follow the rivers all the way to Australia. Now the Bensbach flows into a windswept estuary and disappears into the waters of the shallow Torres Strait.

This strait, 180 kilometers wide, would seem to be a formidable obstacle to human migrations. But in those distant days it did not exist. The rivers of New Guinea watered a land bridge 1,000 kilometers wide that joined New Guinea to the Australian mainland.

The bridge has been present intermittently during the past 2,500,000 years, whenever ice ages caused sea levels to drop. It was especially wide and substantial during much of the past 100,000 years, as the world went through an unusually severe ice age. Then, during the most recent 10,000 years, the world warmed. Ocean levels rose an astonishing 200 meters, separating Australia from New Guinea.

In Chapter 5, we saw how a combination of factors gave our ancestors the technological and cultural tools that enabled them to leave Africa and venture into the rest of the Old World. They were assisted by their temporary resistance to the most dangerous type of malaria, and by the advantage of moving into new environments where new ecological resources awaited them. In this chapter I will explore how some humans managed to spread all the way to Australia, and follow the evolutionary consequences of the long and eventful migration that brought them there.

In the final chapter we will see that modern humans were not the first migrants along this immense pathway. We will ask who preceded us, and why. And we will explore whether the fact that those earlier migrants managed to get almost as far as the later modern human migrants means that we are not as uniquely clever and talented a species as we imagine ourselves to be.

The final stage of the Great Migration

Humans arrived in Australia at the end of a long migration path: from northeastern Africa, through the Middle East and South Asia, down the great peninsula of Southeast Asia, and finally across the treacherous island chains of Indonesia. When the first people arrived in Australia they were venturing into an unimaginably ancient land. They had no idea of the long geological history that they were encountering. But they did find that the land was covered with dense acacia forest and brush. Hidden in the brush were delicious kangaroos and wallabies that were also able to move freely between Australia and New Guinea. One of the first things the new arrivals did was to start burning the brush, to flush out the animals that lived there.

When did the first migrants arrive? Near the coast of Australia's Northern Territory, excavations at a number of sites have revealed stone tools, bits of burned bone, and broken and worn pieces of ocher that had perhaps been used as crayons to decorate people and tools. The first people to arrive had left these bits of evidence on the surface, but they are now buried a meter or more down in the sandy soil.

Figure 105 Australian pelicans, *Pelecanus conspicillatus*, cruise quietly along the green banks of the Bensbach.

In many cases the burned bones are too old to be dated directly using the carbon-14 method, but it is possible to date the sand grains that currently surround them. Or, to be precise, it is possible to date the time at which the sand itself was last exposed to sunlight.

The dating techniques depend on the fact that sand grains contain crystalline materials that gradually accumulate energy, through natural radioactivity and cosmic rays. Electric charges gather at faults in the grains' crystal structure at a steady rate over time. But this clock of accumulating energy is set to zero whenever the sand grains are exposed to sunlight, which blasts away the accumulated charges.

As soon as the sand is covered and away from light, charges can gather again and the clock begins to tick. When samples of sand from the excavations are carefully gathered without exposing them to light, they will glow when heated or irradiated with a pulse of light. The brighter they glow, the longer the sand grains have been buried. Alternatively, the grains can be subject to a strong magnetic field and then bombarded with microwaves that cause the electrons to flip between charged states. This method also measures the strength of the accumulating charges in the sample. It has the pleasing property that unlike the other methods it can be used over and over again on the same material.

In a wide variety of sites around Australia and New Guinea, these techniques and uranium isotope dating have revealed that humans first arrived at least 45,000 years ago, and perhaps as long as 60,000 years.[1] Some more equivocal older sites have been found that may push human history in Australia and New Guinea back to even more than 60,000 years. But the confirmed current oldest data of 45,000 years has not yet been definitively breached.

Still, the time from 45,000 years ago to the present is a huge stretch of history. Napoleon inspired his troops by haranguing them in the shadows of the great monuments of Egyptian civilization, declaiming that forty centuries looked down upon them. Four thousand years is a period of time that encompasses most of Western civilization. But the Cockaded One was boasting about less than 10% of the span of time that humans have inhabited Australia.

What happened during all that time? The data are sketchy, but it appears that the earliest migrants may have entered Australia by crossing a substantial span of open sea from Timor and then spreading out over the rest of the continent. Details are still unclear, but such an entry from the north is supported by the genetics of the present-day Aboriginal tribes. The most genetically variable tribes are found in the north of the continent, and variability is reduced in the populations of the south and west. As we saw in the previous chapter, such a pattern is typical of populations that spread by migration.

Figure 106 (*opposite*) A woman of the Tiwi Islands that lie in the Arafura Sea between Australia and New Guinea. These people, isolated from the majority of the Aboriginal populations, have developed their own culture and striking artistic tradition.

This is because the migrants when they disperse take only a sample of the genes of the founding population with them. But a puzzle remains. The tribes of New Guinea have slightly different mitochondrial DNA from Australian aborigines, suggesting a separate migration across the more northern islands of Indonesia to New Guinea.[2] So why did the Australian and New Guinea tribes apparently not commingle across the great land bridge that connected them? It is possible that there were early migrants from New Guinea into Australia and that they were replaced by later arrivals from Southeast Asia, confusing the story.

Reliving the migration

Although the genetic evidence is lacking, artifacts clearly show that people were moving south through parts of the great land bridge that connected Australia and New Guinea. On the way they built fires, and used ocher pebbles that they found on the stony beach to decorate themselves and to draw pictures and symbols.

One can still get some notion of what their experience might have been like. The Tiwi Islands lie in the Arafura Sea to the west of the Torres Strait and just north of the Australian city of Darwin.[3] During the ice ages they were part of the land bridge that connected New Guinea to Australia. Now the Tiwis are a pair of large flat islands, covered with acacia forest and home to about 2,000 Aboriginal people with their own distinct culture and elegant art.

Even though the Tiwi Islanders are separated by only a narrow channel from the mainland, they are the most genetically distinct of all the Australian tribes. Their fingerprints show significantly different patterns from those of the mainland tribes, and they tend to have a lighter skin color.

One bright winter's day I wandered along the sweeping pebble-strewn beach that fringes the southern shore of Melville Island, the larger of the two Tiwi Islands. The flat coast of the Australian mainland was clearly visible across the narrow Clarence Strait. Among the many colored pebbles on the beach were numerous chunks of ocher, iron oxide mixed with various other

Figure 107 The beach at Melville Island, looking toward the Australian mainland.

minerals. If you pick up these pieces of mineral, dip them in water, and draw them across the skin of your arm, you will leave a vivid red or yellow streak. The Tiwi Islanders use the plentiful ocher for many kinds of decorations.

These pebbles of ocher are the product of a far more ancient set of events than the arrival of modern humans. The gigantic iron deposits of northern Australia date back to 2 billion years ago, when the first free oxygen produced by cyanobacteria combined with soluble iron in the oceans and resulted in huge "red-banded" deposits. These deposits now help to supply the industrial revolution with iron.

Chunks of ocher lay everywhere on the Australian landscape that was encountered by the early human migrants.[4] Some of the pictures that the new arrivals drew with them are preserved in nearby Kakadu National Park. The pictures include drawings of the striped catlike thylacine marsupial known as the Tasmanian Tiger or Wolf. As we saw in Chapter 3, this predator was finally driven to extinction in the 1930s in its last refuge, the isolated southern island of Tasmania. But it went extinct on the Australian mainland about 3,000 years earlier. It is unclear whether it was the arrival of the wild dingo dog that triggered the extinction of the thylacines and other large marsupials around that time, or whether the extinction was due to improvements in Aboriginal hunting technology that have been dated to about the same time. But the marsupial tigers must have been plentiful when the first human migrants arrived in Australia.

The first humans did not arrive empty-handed. They brought fire, weapons for hunting and war, and possibly tales and legends from their earlier travels. Once they began to spread in Australia their languages diversified and so did the legends that they told. Present-day northern Australian legends and beliefs provide some clues to the legends imported by their ancestors, though much has changed in 45,000 years.

One powerful deity is the Rainbow Serpent, a god of water. He used his sinuous form to push up mountains and ridges, and as he writhed he carved out the winding courses of rivers. Some of the pictorial representations of the Rainbow Serpent are as much as 6,000 years old, and he continues to figure in present-day Aboriginal ceremonies. The worship of the Rainbow Serpent must be the oldest surviving religion on the planet.

The Rainbow Serpent deals with water in its various manifestations, so it is not surprising that he plays a more important role among the northern tribes than among those in the dryer south and west. Did some ancestral form of the Rainbow Serpent legends accompany the early migrants across the land bridge from the even wetter north, more than 40,000 years ago? If so, then this religion would be ten times older than any other religion that we know of.

Australia and Tasmania were the furthest destinations that these human migrants could reach. They were finally stopped in their great trek by the

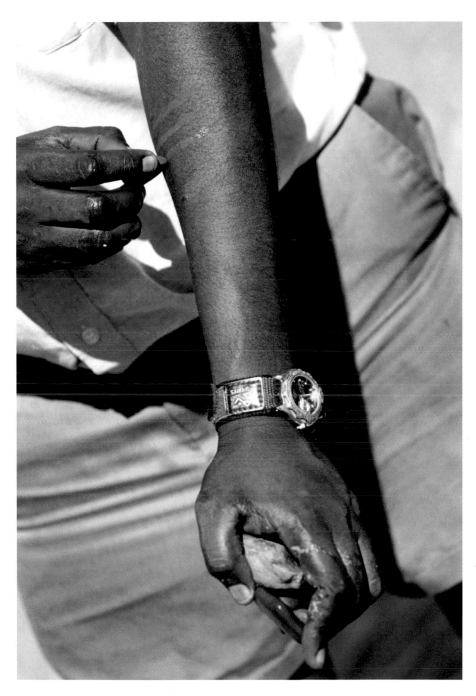

Figure 108 Chunks of ocher, in this case from the Melville Island beach, leaves vivid red, orange, or brown stripes on the skin when they are wetted and used as crayons. There is archeological evidence that the earliest Australians used ocher for decoration.

Figure 109 You can make out the distinct stripes on the rear end of this petroglyph representation of a Tasmanian tiger at Kakadu National Park. The petroglyph, protected by a rock overhang, must be at least 3,000 years old, because this is when these marsupial predators disappeared from northern Australia.

impassable southern ocean. New Zealand lay beyond, but it was separated from Australia by 3,000 kilometers of treacherous water. Humans would not arrive there, by island-hopping from the north, until 750 years ago.

The Great Migration's route out of Africa

The spread of peoples from New Guinea through Australia and finally to Tasmania was the last stage of this most extensive of all early human journeys. From the evidence that we have, this migration penetrated even further into the unknown than the migration of *Homo erectus* from Africa into Asia

Figure 110 This Rainbow Serpent petroglyph from Kakadu may provide a link to the religious beliefs of the people who first migrated to Australia.

almost 2 million years earlier. *Homo erectus* seems to have stopped in Indonesia. *Homo sapiens* went on to New Guinea and Australia.

The modern human journey, like the earlier migration of *H. erectus*, probably started in Ethiopia, and it started hesitantly. Fossil finds by Tim White and his colleagues show that people much like us were living in Ethiopia 150,000 years ago. And skeletons discovered in the 1930s in Skhul and Qafzeh caves on the slopes of northern Israel's Mount Carmel show that modern humans had arrived there by 90,000 years ago.

But genetic evidence shows clearly that it was exceedingly hard for these small groups of modern humans to move away from the Mediterranean coast. While there are many different mitochondrial lineages in human populations in Africa, the populations that first managed to penetrate beyond

Figure 111 The challenges facing early human migrants as they traversed the blistering coasts of what is now Iran and Pakistan must have been considerable, but their ingenuity was clearly up to the task. Even at the present time, in the extremely unfriendly desert of Sinai, porous dyke intrusions lace the hills and bring water to unexpected places. The walled Bedouin garden seen here has been maintained by these water seeps for hundreds of years.

the Middle East into southern and northern Asia carried only one of these lineages. Stephen Oppenheimer of Oxford University and his co-workers suggest that this is evidence that only one small group made it through the deserts to the east of the Mediterranean. They point out that other notable human migrations such as the invasion of the Americas carried several mitochondrial lineages with them.

Why was the breakout from Africa and the Levantine coast of the Mediterranean so difficult? The logical route for these early migrants to take would have been north and then east, through what were then the wet and

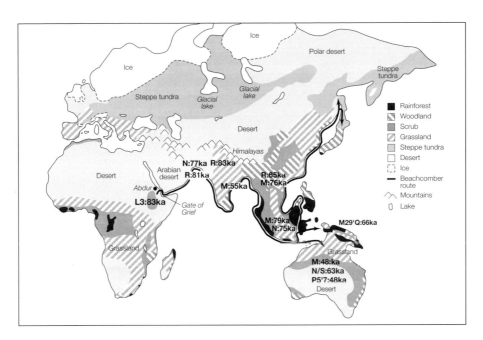

Figure 112 Stephen Oppenheimer's map of the Great Migration, including a possible later extension to the north toward China and Japan. The letters and numbers refer to the ancient mitochondrial lineages carried by the various groups and the approximate times when they arose: for example, lineage L3:83ka arose in Africa approximately 83,000 years ago. I suspect that the beginnings of the migration followed a more northerly route than the one shown here that skirts the southern coast of Saudi Arabia. That coast, currently extremely desiccated, must have been even more forbidding during the dry climate of the Ice Ages. (From Stephen Oppenheimer, "The Great Arc of Dispersal of Modern Humans: Africa to Australia," *Quaternary International* 202 (2009): **2–13**, Figure 3.) © Stephen Oppenheimer.

fertile regions of eastern Anatolia and northern Syria, and finally down into the lush country of the Tigris and Euphrates river valleys. But these early migrants out of Africa must have encountered well-settled populations of Neanderthals that had been living in Syria and as far east as Iran for millennia.[5]

The conflicts that might have taken place between these two groups of peoples are now lost to history. But I suspect that the Neanderthals were anything but pushovers. We know that they made axes with stone blades and wooden hafts, and buried flowers along with their dead.[6] We know nothing about their martial skills, but we do know that they had survived for a long time in the Middle East before modern humans arrived.

It is striking that, during the period between 50,000 and 100,000 years ago, only the remains of Neanderthals and no unequivocal traces of modern humans have been found in the whole vast region to the north and west of the Tigris–Euphrates valley. The Neanderthals seem to have disappeared from this region about 50,000 years ago. And we know that modern humans must have passed through Neanderthal territory, or somehow dodged around it, during that time. The mitochondrial DNA story provides strong evidence that only one small group of modern humans made it through.

Once the modern humans managed to get past this formidable barrier, the logical route to follow would have been down the Tigris and Euphrates, then along what are now the coasts of Iran and Pakistan to the mouth of the Indus. This route, much of it along a dry, rocky, and forbidding coast, must have posed severe challenges, but any more northerly route would have been even more difficult. We do not know when they accomplished this feat, except that it had to be at least 50,000 years ago. Once the Indus was reached, it was relatively easy to travel down India's verdant west coast to its southern tip.

The extent of the Great Migration

Some groups from this Great Migration stayed behind as others, more adventurous or more foolhardy, moved on. We find descendants of some of these populations that stayed behind still living in southern India and in the northern Andaman Islands.

Southern India is a palimpsest of tribal groups and identities, and although anthropologists have long known that this mix of different human types had a long history, it was not until the advent of mitochondrial DNA and whole-genome analysis that it became clear how long and complex this history really is.[7]

Although many members of southern India's aboriginal groups have recently moved to the cities, some have stayed behind in remote regions to which they had been driven by later waves of migrants. One of these is the Nilgiri Hills, the dramatically isolated mountainous plateau that lies at the conjunction of the states of Tamil Nadu, Kerala, and Karnataka. We encountered this plateau, which lies to the east of Mudumalai National Park, in the introduction to this book. The top of the plateau is a refuge for ancient tribes such as the Bagada, Toda, Kuruba, and others, who have retreated there from the coasts.

At Mudumalai's research station I met some of the rangers who watch over the plot and who belong to the Kuruba aboriginal tribal group. These intelligent and enthusiastic workers are excited about being given the responsibility of overseeing a large scientific project.

The Kuruba are known across southern India for their cross-dressing festivals in which men dress up as women and make raucous fun of any upper caste people who pass by. They are more acculturated than the shy Toda, who live higher in the hills and have been reduced in numbers to only about 1,400. The cattle-herding Toda live in tiny thatched houses that look like miniature Quonset huts. The women, unlike the women of any other tribal group in the area, take several husbands.

So much cultural mixing has taken place in this part of India that it is impossible to discern which traditions go back the furthest. Luckily for our understanding of their history, these tribes and other South Indian groups still carry a few copies of some of the non-African world's most ancient types of mitochondrial DNA. Some branches of their mitochondrial family tree stretch back to the time of the Great Migration.[8]

Equally ancient DNA lineages are found elsewhere in India and at other points along the route of the Great Migration. One of these is the Andaman and Nicobar Islands, part of a chain of islands that extends south from the coast of Myanmar into the Indian Ocean. These islands were thrust up by the great collision between the Indo-Australian and Southeast Asian tectonic plates.

The Negrito people who inhabit the Andamans and Nicobars have physical characteristics that resemble those of some African populations. They are as short as African pygmies. The inhabitants of the more acculturated islands to the south have been driven to near-extinction by a combination of introduced diseases and careless colonial administration, first by the British and later by the Indian government after independence. But the tribes on the northern islands have managed to stay isolated. The tribe living on North Sentinel Island in particular has fiercely rejected all attempts at contact from the outside, including fighting off intruders with spears. In view of the Andamans' history this seems like a sensible attitude.

The Andaman Islanders were clearly part of the Great Migration. Analysis of their mitochondrial DNA has uncovered sequences with deep roots, like some of those carried by the tribal populations of India's southernmost tip. These roots can also be traced back to Africa.

The islanders probably did not originate from India's southern tip, but from later migrants and settlers who continued to move slowly up the eastern Indian coast and through the vast swampy delta of the Ganges-Brahmaputra, finally reaching the Irrawaddy. These new arrivals probably did not settle the islands immediately, for there was plenty of room on the mainland and there would have been 400 kilometers of open ocean to cross.

About 20,000 years ago, during the most severe of the recent ice ages, it might have been possible to walk most of the way to the islands along the now-submerged Coco Island ridge. But there was no point at which the islands were linked to the mainland by a continuous land bridge. The islands were probably colonized by people in small coastal boats, some of which may have been blown out to the islands by storms.

The saga continues

Bands of hardy hunter-gatherers and fishermen continued to spread, across the delta of the Irrawaddy and down the extensive coastal marshes of peninsular Southeast Asia. Beyond the end of the present-day peninsula it was easy to traverse most of the islands of the Sunda Arc, because 50,000 years ago sea levels were so low that they were joined together. Sumatra, Borneo, Sulawesi, and Java were all linked by the Sunda Shelf, a great expanse of flat and friendly land that must have had abundant fish, game, and water. But in order to reach New Guinea and Australia these migrants still had to make their way across relatively narrow, deep ocean channels that were swept by strong north-to-south currents.

The first of these channels was the deep-water passage between Borneo and Sulawesi. A narrower and much shallower passage separates Bali and Lombok to the south, but it may have been even more difficult to cross because anyone venturing out in a primitive boat or accidentally carried out to sea would have been swept by the strong currents to almost certain lonely death in the Indian Ocean to the south. Other formidable channels separated islands further to the east, cutting off New Guinea and Australia from the island arca that is now known as Wallacca.

Michael Morwood of Australia's University of New England suggests that the migrants may have been able to make the crossings most easily along the northerly route from Borneo to Sulawesi and then through the Spice Islands to New Guinea, where the currents were less likely to sweep them south and out to sea before they could make landfall. Regardless of their route, the migrants had to make several sea passages as they moved further east. They might have possessed the seafaring technology to brave the swift currents, or the transits of the passages might have been as simple as an accident of weather—a coastal boat, laden with men, women, and children, blown out to sea by a sudden storm. When they managed to reach Australia and New Guinea they once again found themselves on a vast continent-sized region that awaited exploitation.

Throughout all these population movements the early migrants relied on a warm climate, plenty of fish and game, and the absence of truly daunting physical barriers. They were unlikely to have traveled very far inland at first, but soon population pressure and the disappearance of animals as they were over-hunted drove them to exploit areas away from the coasts.

We know something about the Southeast Asia migrants from some of their descendants, groups of hunter-gatherers who live in the dense rainforests of the Malay Peninsula.

The lowland rainforests of the peninsula covered vast areas throughout the Age of Mammals. Although individual forests have come and gone, there have always been substantial regions of forest on both the peninsula and the Greater Sunda Shelf. These forests are by many measures the most diverse on the planet. Dense and deeply shaded, they are dominated by immense *Shorea* and other Dipterocarp trees. Giant jackfruit and other fruit trees provide rich resources for the orangutans and other animals that live in the high canopy.

The forests of Southeast Asia are magical places. Often the trees flower all at once, a phenomenon called masting that is triggered by just the right combination of temperature and moisture. After this massive explosion of flowers in the forest's upper canopy, the seeds mature. The Dipterocarp pods split into halves and shower down to the forest floor far below like swarms of tiny helicopters—hence the name Dipterocarp, which means two-winged fruit. This synchronization overwhelms with its sheer abundance the animals that eat the seeds.

The small groups of fishermen and hunters who were the vanguard of the Great Migration encountered many animals in these forests. The small Sumatran rhinoceros and the Sumatran tiger, both now almost extinct, were the most dangerous. Great troops of langur and proboscis monkeys, gibbons, and macaques swung through the trees. Orangutans ranged widely through the forests, leading their solitary lives high in the canopies. Wreathed and rhinoceros hornbills flew overhead, their wings sounding like the thrum of helicopter blades. And of course the swarms of smaller life, from birds to insects, made the forests vibrate with noise and activity.

The environment that these adventurous migrants stumbled into was more varied than a featureless wall of rainforest. In the early nineteenth century extensive rainforest covered 90% of peninsular Malaysia, but it probably covered less than half of that during the maxima of the ice ages when there were alternating wet and dry seasons instead of the year-round rains of the present time. The drier forests of the ice ages were interspersed with open grassland and areas of low shrubs. This open country must have given the early hunters many opportunities to track down the abundant tapirs and pigs.

The first human hunters who arrived in this dangerous but abundant Eden found what seemed to be infinite resources. And their descendants are still exploiting them. The forest people of Malaysia continue to hunt intensively, as they have for tens of thousands of years, even though the rainforests are sadly diminished in extent and most of their animals have already been killed.

Rubber trees and oil palms have replaced most of the rainforest. A flight over the Malaysian lowlands today reveals an unnervingly uniform landscape. Gone are the infinite shades of green of the lowland forest, where the crown dipterocarps towered far above their neighbors. Instead the landscape resembles an immense pair of green corduroy trousers, stretching out to the horizon. The corduroy is made up of row upon row of oil palms.

Oil palms grow rapidly, producing their first crop of fruit in three years. In contrast even a planted teak forest, which despite its uniformity does retain some of the original forest's diversity in its understory, will take 70 years to mature. It is obvious which type of crop is more likely to be funded by banks and corporations impatient for a return on their investment. The replacement of the Malaysian forests by oil palms and rubber trees has greatly increased the pressure put on the remaining fragments of forest by the hunting activities of the aboriginal population.

There are some rays of hope. Discoveries over the past decade of a cornucopia of new animals and plants in peninsular Southeast Asia and Borneo sparked news stories around the world. In February 2007, the three countries that have divided up Borneo agreed to protect about a third of it. This "Heart of Borneo" agreement has been brokered by the World Wildlife Fund and the government of Brunei, and there is growing enthusiasm for the agreement on Borneo itself. It is the largest ecological treaty in human history, involving at least 220,000 square kilometers. By comparison, Yellowstone Park in the western USA protects only 8,980 square kilometers.

Unlike Borneo, most of peninsular Malaysia's ecological diversity has already been destroyed, and there is only one official national park. This park, eponymously called Taman Negara or "national park," along with some other protected areas, account for about 4% of the originally forested area.

The aboriginal peoples of the region, known as the Orang Asli or "first people," carry the genetic legacy of the scattered bands of the Great Migration who trickled through peninsular Malaysia and fanned out across the open plains of Sundaland. Like the earliest settlers of southern India, these tribes have intermingled with later arrivals. They were probably driven from their first settlements on the coasts into the interior forests by the spread

Figure 113 Peeling the "wings" off dipterocarp seeds in Sabah's Danum Valley. The seedlings of these giant rainforest trees will be planted in experimental forest plots.

of later seafaring invaders. The Orang Asli have been persecuted and even enslaved in the past, but despite this grim history they have played an important role in the history of modern Malaysia and its fight for independence.

These aboriginal peoples have remarkably diverse cultures, because they have had a long history of isolation. The extreme cultural diversity found in New Guinea today, with its 800 different languages, is the result of New Guinea's precipitous and fragmented geography. The Orang Asli live in a less fragmented world, but they are still currently divided into nineteen distinct language groups and their ancestors undoubtedly spoke even more languages.

I have visited only one of these groups in peninsular Malaysia. The members of the Batek tribe live on the periphery of Taman Negara. They are allowed to hunt with blowpipes and bows and arrows in the park itself, which is part of their ancestral hunting ground. They live in palm-thatch settlements along the three rivers that drain the park's more that 4,000 square kilometers. Some of them work as guides, but most simply move quietly in

Figure 114 These Batek tribespeople live on the edge of Taman Negara National Park in peninsular Malaysia. The Batek, descendants of the first modern humans to traverse the peninsula, are permitted to hunt in the park.

and out of the forest, growing yams and hunting monkeys and squirrels. The children receive little schooling, and run around cheerfully all day.

The Batek are smaller than the average Malay and their skin is darker. They and the other Orang Asli are not as African-looking as the far more isolated Andaman Islanders, because of a long history of intermarriage with neighboring tribes. But despite this genetic mixing some of them still carry the deep-rooted mitochondrial DNA sequences that have been passed down from the time of the Great Migration.

In Borneo the aboriginal diversity is even greater. As of this writing the sets of complete mitochondrial sequences essential for tracing the Great

Figure 115 Kayan woman from the Mulu River area of central Borneo.

Migration are still being collected from Bornean tribespeople. But they are likely to be as deeply rooted as the Orang Asli—we will see in the next chapter that the area around Niah Cave near the coast in Sarawak was settled at least 40,000 years ago.

I have visited the Kayan of central Borneo, known as the "long earlobe" people, who live on the banks of the Mulu and other rivers of the interior. And I spent a delightful night as a guest in a longhouse of the Rungus people of northern Sabah, near the northernmost tip of Borneo. The Rungus preserve some of the oldest traditions in Sabah, but they are clearly Malaysian in appearance. It is obvious that any genetic and cultural connections between them and the early migrants from Africa are tenuous at best, diluted by generations of intermarriage and the adoption of new customs.

In the Mulu River area of central Sarawak the results of intensive hunting are clear. Sarawak is known as the Land of Hornbills. Alas, hornbills are few and far between now, their populations decimated by native hunters and by

Figure 116 Rungus longhouse, northern Borneo.

the guns of poachers from the peninsula looking for skins and feathers. They can occasionally be seen in the distance, flying above the trees. And other animals are in short supply as well. In a day-long hike near the immense Mulu caves I managed to spot one small and extremely worried-looking squirrel. As we saw in Chapter 5, the removal of top-predator animals and birds from these forests will have a devastating long-term impact on their diversity even when the forests themselves remain intact.

The dynamics of the Great Migration

The Great Migration was not some inexorable march of peoples determined to reach Australia. It was a far slower and more diffuse process. Small groups of people fanned out into unpopulated country, establishing encampments and adapting to the new conditions that they found. Many of these tribal groups, isolated for long periods, developed unique languages and cultures.

Figure 117 Rungus woman playing nose flute, northern Borneo.

It was not long before these diverse tribes began to war with each other, as tribes still often do in New Guinea. And rapid exhaustion of the easiest prey and pressure from population growth continued to push the migrants into new lands. The vanguard of the Great Migration was likely to have been a hesitant one, mostly made up of tribes that had been forced from their lands when they faced food shortages or lost skirmishes with other groups.

We can now get some notion of how long it took for the Great Migration to happen. Using rapid DNA sequencing technology, it is now possible to sequence complete mitochondrial chromosomes from the peoples scattered along the Great Migration route, and to look for deep branches in the resulting family trees. These exhaustive data were recently brought together, organized and augmented by Vincent Macauley of the University of Glasgow and his colleagues. They found a remarkable pattern.

All the populations strung out along the route of the Great Migration carry descendants of a mitochondrial chromosome, L3, that came from Africa. And all the populations carry a mix of two descendant types of chromosome, M and N. DNA base changes that accumulated in the M and N chromosomes, however, are different in the different populations.

The differentiation into M and N took place about 65,000 years ago. Within each of these sequence types there were further divergences, but these additional divergences are unique to each population. Such a pattern could only have happened if the original migrants had moved rapidly along the migration route and the populations that they left behind remained isolated from each other, free to diverge in different directions. Once the colonies were established, their mitochondrial chromosomes diverged independently of each other, showing that the groups did not mix after the initial migration.

The data also showed that the Great Migration really did take place along the southern rim of Asia, and did not branch off early to the north as some paleontologists had proposed. The aboriginal peoples who first occupied regions that lie to the north of this route, in Taiwan and on the eastern Asian mainland, carry samples of mitochondrial chromosomes that began to diverge from each other more recently than at the time of the Great Migration.

Of course, that 65,000-year date has a large error associated with it. The authors suggest an error that could be as small as 3,000 or as great as 12,000 years. On the basis of this error, they estimate that the migrants could have covered the entire 12,000-kilometer route of the Great Migration, from Africa to Tasmania, at an average rate of between 1 and 4 kilometers per year.

At the slower rate, the migration would have taken 12,000 years, and at the faster rate only 3,000. Obviously, whether the migrants moved quickly or slowly, they did not travel at a constant rate. They must have leapfrogged quickly along the more forbidding stretches of coast and dawdled in the lusher regions. And even the shorter time span would have provided numerous opportunities for tribal warfare. These wars and the great distances involved split the migrants into factions.

Figure 118 Scaffolds used by poachers of birds' nests dot the immense Niah Caves of Sarawak. These caves and the nearby areas have been inhabited by modern humans for at least 40,000 years.

Encounters along the way

As we saw, the modern human migrants must have encountered Neanderthals in the Middle East, but traces of those encounters have been lost. Later, when their descendants on the Great Migration arrived in the Greater Sunda Islands, they encountered some even more distant relatives.

The first of those relatives to arrive on the Greater Sunda island chain were (we think) *Homo erectus*. Bands of *H. erectus*, our highly peripatetic and modestly brainy ancestors, left Africa about 1.8 million years ago. During the early part of their own extensive migration they appear to have followed migration routes far to the north of the later modern human Great Migration. They left fossil traces behind in Georgia, north of Anatolia. But once they reached

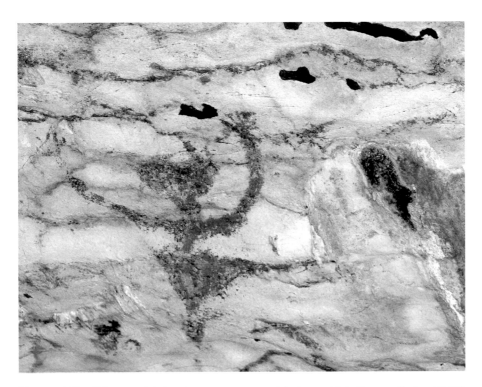

Figure 119 This graceful dancing woman was painted on a Niah Cave wall about 2,500 years ago. Humans have inhabited these caves for 40,000 years, but it is not known when they first began to paint the walls.

Southeast Asia they must have followed a route similar to that of the later modern humans. Eventually they arrived on Java about 1.5 million years ago.

Astonishingly, it appears that *H. erectus* survived on Java for an enormous span of time, down to almost the present day. Between 1994 and 1996, Berkeley geochronologist Carl Swisher and his colleagues were able to get accurate dates of bovine teeth found in the same riverbank strata as the *H. erectus* bones, along with dates on bits of volcanic deposits trapped in the skullcaps themselves. Their measurements yielded values ranging from more than 1.5 million years ago to as recently as 28,000 years ago.[9]

If *H. erectus* really did survive on Java for all this time, longer than the entire history of hominids in Europe, then it seems likely that they were still there when modern humans first arrived. Did *H. erectus* and modern humans actually encounter each other on Java?

Almost certainly yes. The oldest incontrovertible traces of modern humans on Java, in a cave called Song Gupuh in the eastern part of the island, have been dated to only 12,000 years ago.[10] But there is plentiful evidence from other nearby islands that modern humans arrived in the region at least 40,000, and perhaps as much as 60,000, years ago. Thus they could easily have overlapped with the last of the *H. erectus*.

The world that modern humans found as they ventured into Sundaland must have been strange and different from anything that they had encountered earlier. I got a faint idea of the conditions that they must have encountered when I visited the vast series of caves in Niah National Park. This rainforest park is an ornament of the Malaysian province of Sarawak on Borneo. I went there with my colleague Sylvester Tan, a Bornean forest ecologist whom you met in Chapter 5.

The heavily forested Niah Mountains seem solid enough, but in fact they are honeycombed with gigantic caverns. The caves are some of the largest in the world, immense hollows that have been eroded out of the soft limestone of the park's mountains by millennia of rainwater. Here and there the remaining thin limestone shell that separates the caves from the outside world has collapsed, admitting shafts of green light into the immense spaces within the mountains.

These caves, as large as aircraft hangars, have a sadly deserted feel. Once they were the home of colonies of millions of swifts that built their nests of twigs and spittle on the caves' roofs. The colonies have been decimated to feed China's insatiable appetite for birds' nest soup. When I scanned the caves' roofs I saw very few nests.

Graeme Barker of the University of Leicester and his colleagues have shown that the caves and the area around them were first inhabited by modern humans at least 40,000 years ago.[11] The clinching find was a modern human skull dating from that time.

The first modern humans who arrived at the caves must have found them filled with swifts and bats. Indeed, a few days earlier I had watched at sunset as immense swarms of bats burst out of the nearby Mulu caves. These caves, almost as large and extensive as the Niah caves, are in Sarawak's forested interior near the wild and beautiful Mulu River. They have not been as damaged by poachers because they are more remote.

The bat populations of the Mulu caves have also undergone steep recent declines in numbers, but they are still impressive and provide some indication of the abundance that must have greeted the first modern humans in the area. Seemingly limitless food sources must have encouraged these early migrants to settle. Then, overtaken by restlessness or more likely forced to move on after they had overexploited the easiest supplies of food, some of these people traveled further south through Borneo's thick jungles and across the land bridge that then joined Borneo to Java. During this migration they must have encountered ancient tribes of *H. erectus*. Did they fight with them? So far the fossil record is silent on any interactions.

At the end of the migration

The genetic evidence is strong that modern humans did migrate over the space of a few millennia across the enormous distance from northeastern Africa all the way to Australia. How could they, with their Stone Age technology, move so far so quickly? Some answers can be found by examining what we know about the history of the peoples with the clearest cultural connection to the original migrants, the Australian Aborigines.

The Aboriginal tribal land of Arnhem Land in Australia's Northern Territory has close associations with the first migrants who came over the land bridge from the north. On a 2001 visit, before the full details of the Great Migration had become evident to genetic anthropologists, I had the chance to penetrate this remote area.

The tribal reserve covers a vast region from Kakadu National Park all the way east to Groote Eylandt in the Gulf of Carpentaria. Australian tribal lands have had an agonizing history, beginning with the first violent clashes over territory in 1788 between the Aboriginals and the Europeans who had begun to spread out from the convict settlement at Botany Bay. The clashes continued over more than two centuries, marked by continuous conflict

Figure 120 Dawn breaks in an Arnhem Land tree fern forest. The first migrants to Australia encountered similar landscapes.

and the broken promises made by dozens of successive national and state governments. Now there are some signs that the territorial conflict might finally be resolved. In 1992 the Australian High Court admitted that Aboriginals have a right to their tribal lands. The court also admitted, startlingly, that Australia had never actually been an "empty land," totally open to European colonization and exploitation, as the first European settlers had proclaimed. But even after the court's decision, argument continues over the extent of tribal lands and about who is entitled to live in them. The result has been a menagerie of government commissions that continue to spar with Aboriginal tribal councils.

A small group of us, including Aboriginal guides, was headed across the Arnhem Land tribal territory toward the shores of the Gulf of Carpentaria. We camped along the way, cooking damper bread and bathing when we found a stream. Our Land Cruiser encountered endless difficulty because of the recent rains, and we were repeatedly bogged down on the muddy track.

One morning one of the guides found and killed a common brown snake, which is unusually dangerous despite its proletarian name. We were all, as a consequence, nervous as we penetrated deeper into this lonely land.

In the early mornings the spiderwebs that festoon the tree fern forests of Arnhem Land are covered with dew. The ferns, stumpy and closely spaced, dominate a green and gold landscape. Apart from the grasses and the occasional eucalyptus tree, it could be a world before the time of flowering plants. Indeed, it looks as if a dinosaur might blunder through it at any moment. As the sun rises and the dew burns off, the landscape is revealed in greater detail, alien and still strangely silent.

Arnhem Land covers about 75,000 square kilometers, and is home to only about 16,000 tribespeople. One can drive for hours along the muddy tracks and see no sign of human occupation. We did stumble on an encampment of sketchy houses and lean-tos that had been abandoned days earlier when the people who lived in them had disappeared into the bush.

The sparseness of human settlement was paralleled by a lack of wildlife, especially compared with the thriving birds and animals of Kakadu National Park that lay immediately to the west. In five days we saw a few feral water buffalo, a dingo, and a solitary tawny frogmouth that was perched on a tree and turned its dazzled yellow gaze toward our headlamps.

As we headed across the drier plains on the southern leg of our trip one of the reasons for the lack of wildlife became clear. Our guides kindled branches and tossed them from the truck, leaving a series of blazes behind us. I protested, saying that I felt like Smokey the Bear in reverse. But I was ignored: the new generation of young people living in Arnhem Land tends to set fires at seeming random, with no notion of how far or how fast the fires might spread.

Fires have always been part of Australia's ecology, but the paleontological record shows that layers of charcoal from widespread burning increased in thickness and in frequency with the arrival of humans 45,000 years ago. The first Europeans to arrive in southeastern Australia were astonished by the mastery of fire shown by the Aborigines. They set controlled fires, each covering an acre or so, in a patchwork pattern that changed each year. Then, before the fires got out of control, groups of people deployed swiftly and beat them out using green branches. On average, a piece of land was burned off about once every five years.

This controlled burning had a huge effect on both the landscape and the plant life. Just as the Indians of the American Northeast used fire to open up the landscape and make it easier to hunt, the effect of the Aborigines' burning was to open up the wetter regions of the Australian landscape. The result was extensive open grassland with a scattering of large trees. As soon as the Europeans began to fence off the land and raise sheep and cattle, brush and small trees were free to grow up. As in the American West, where the summers are dry, this ecological change has exposed entire ecosystems to disastrous wildfires that now threaten entire cities.

Because the Aborigines had always been hunter-gatherers they depended on an abundance of wild plants. Before the arrival of Europeans a great variety of seeds and tubers accounted for at least half their diet. In southeastern Australia one of the most important of these is the Murnong plant, a tough desert species of the Aster family.[12] Murnong has a nourishing tuberous root, and thrives in areas recently swept by fire. Nineteenth-century settlers wrote of seeing huge fields made up of millions of these wild plants. Large numbers of Aborigines would wander among them, digging up the roots. The Aborigines had not deliberately planted the Murnong, and thus had not invented agriculture in the formal sense. But they were perhaps on their way to an agricultural revolution before they were interrupted by the arrival of the Europeans.

The productivity of grasslands, too, was fire-dependent. In the absence of fire Australian native grasses form dense tussocks with no room for other seed-bearers or tuberous plants. Fire releases the productivity of these lands, but when fire is suppressed in order to protect grazing, tussocks form and the native plants that are unable to survive in them disappear.

The fields of Murnong plants largely vanished with the introduction of sheep and cattle, driving many Aboriginal tribes to near-extinction along with them. Grazing brought negative changes in other parts of Australia as well, dramatically reducing the diversity of plants on which the native peoples depended.

Figure 121 Sunset along Arnhem Land's Bulman River.

The first Europeans remarked on how healthy, strong, and tall the Aborigines were. The survival of the native populations, who had managed to thrive in that forbidding landscape for at least 400 centuries, was the result of a delicate ecological balance in which a diverse mix of wild plants was maintained in part by periodic burning. Disturbance of that pattern has been a disaster, both for the people and for their culture.

The random burning I saw in Arnhem Land was the worst possible combination of behaviors. The young Aborigines who accompanied us, like youngsters everywhere, loved to set fires. Their Arnhem Land elders know better than to start fires at random. Instead, they practice controlled burning to encourage the growth of native plants. In this dry area it is seeds rather than tubers that form a staple part of the diet, and they know that controlled fire at regular intervals encourages a diversity of seed-bearing plants to thrive.

This long cultural tradition of controlled burning may now be gaining new life, ironically through the impact of the modern world. In 2007, to expedite the permitting process for a huge new refinery in nearby Darwin, ConocoPhillips agreed to pay a million Australian dollars to the Arnhem Land tribal councils. The money was earmarked to encourage traditional burning patterns. The idea is to bring under control the immense wildfires (most of them deliberately set) that currently burn off half of Arnhem Land every year, allowing the refinery company to offset its own carbon emissions. We can only hope that this encouragement of traditional practices will help to restore the damaged ecosystem of Arnhem Land and bring back some of the vanished animals and birds.

How do my experiences in this fragile northern Australian world bear on the Great Migration? Do these present-day collisions between tradition and the modern world really have any bearing on how the migration took place and why it was so rapid?

The humans who first migrated out of Africa and along the southern rim of Asia were confronted with utterly new situations, unlike anything their ancestors had encountered during previous stages in their migration. They found extensive coastal mangrove forests along the shores of southern India and Southeast Asia that would have been impossible to traverse except by boat. And they encountered, one after another, the vast estuaries of the great Asian rivers—the Indus, the Ganges-Brahmaputra, the Irrawaddy, and the Mekong. Each river was different, each flowed through different country, and the swamps of their estuaries harbored a different mix of food plants, animals, and diseases that modern humans were encountering for the first time. When the migrants crossed Wallace's Line they encountered an even more dramatic change in the animals and plants. All these new environments posed a continuous series of life-or-death challenges.

My guess is that, just as with the burning of Arnhem Land, similar careless overexploitation of new ecosystems also took place in the distant past during the Great Migration as the migrants moved into new areas. The devastating ecological consequences of this behavior forced some of the migrants to move on, perhaps only a generation or two later. Meanwhile, the people who stayed behind were forced to learn to modify their behaviors in ways that helped to preserve the land's productivity.

How people were changed by the Great Migration

What evolutionary impact did the Great Migration have on the populations that took part in it? To begin with, the migrants remained surprisingly isolated for many thousands of years. There are no signs in their mitochondrial DNA that they mated with the much earlier migrants, *Homo erectus* and their relatives, that they encountered along the route. But such matings cannot be ruled out, as is shown by recent discoveries of a Neanderthal signal in the nuclear DNA of modern humans.

Perhaps the biggest puzzle is why this migration seems to have been a unique event, bringing with it only a small sample of the genetic diversity found among modern human populations in Africa. The genetic pattern seen along the southern rim of Asia suggests that only one band or a few bands of people managed to pass through the barrier of the Middle East, and it was their descendants that migrated all the way to Australia. Why did nobody else follow for so long?

Conditions along the route of the Great Migration may simply have been too challenging, especially in the difficult country of the Middle East. Much of the Middle East would have been desert, and the parts that could support hunter-gatherer populations were already inhabited by Neanderthals. Cores from Antarctic glaciers with ice layers from 60,000 years ago contain an unusually large amount of red dust, signs of dust storms that extended from pole to pole. We are seeing similar worldwide dusty conditions today, as a result of our rapidly spreading deserts.

Nonetheless, people survived in these grim regions over long spans of time. Iranian archeologists have recently discovered massive numbers of stone tools near the city of Semnan in northern Iran. The tools date from

200,000 years ago right down to 40,000 years ago, showing that this area was inhabited since long before the Great Migration began. It seems likely that the earliest inhabitants were Neanderthals, but when and how were they subsequently displaced by modern humans? Puzzling questions of origins also confront us at a living site excavated by workers from the University of Sheffield at Riwat in northern Pakistan the 1990s.[13] This collection of stone tools, along with some tantalizing indications of post-holes that might have been supports for shelters, has been dated to at least 45,000 years ago. Who were the people who left these traces? Were they in transit or permanent settlers? Were they part of the Great Migration? We simply do not know.

At the time of the Great Migration much of the Persian Gulf was land instead of sea, so that the confluence of the Tigris and Euphrates flowed directly into the Gulf of Oman and the Arabian Sea. In theory the lengthened rivers would have ended in a rich estuary that would have provided the migrants with everything they needed.

But when the Persian Gulf was dry, becoming a hot parched Persian Valley, it effectively doubled the length of the Tigris–Euphrates river system before it could reach the sea. The lowered rainfall during this glacial period would have made the rivers smaller and more seasonal, and the hot sun would have beaten down unmercifully on the waters as they flowed through the Persian Valley for that extra thousand kilometers. It seems possible that the Tigris–Euphrates might have faded away, disappearing into the sand, just as the Okavango River disappears into the Kalahari Desert in southern Africa today. When the early migrants reached the end of these rivers they would have been faced with a forbidding barrier of sand followed by a full 2,500 kilometers of dry and inhospitable coast before they reached the Indus Valley at what is now the border between Pakistan and India.

The coasts of Iran and western Pakistan are grim today, and they may have been even grimmer then. If the migrants had to surmount this vast unfriendly region of coast, it would not be surprising if they only managed to do it once. How did they accomplish this feat? They might have crept along the coast, surviving in occasional oases. They might have swung inland, leaving behind some of the stone tools that have been found in Iran and Pakistan. Or they might have leapfrogged the coast by embarking on dauntingly long sea voyages. We have no idea of their shipbuilding skills, but we do know that they carried only the most primitive stone tool technology with them from Africa.

This part of the Great Migration was so challenging that few people made it through. But when those lucky survivors stumbled on the lush estuary of the Indus, they were home free. The eastern coastline of India, green and subtropical, stretched away invitingly to the south. It is striking that Mumbai was home to hippos as recently as 20,000 years ago, even though hippos had vanished from Iran and Pakistan at least a million years earlier.[14]

The way was now clear for the descendants of these tough migrants to spread rapidly. Within a few thousand years some of them made it all the way to Australia.

The Great Migration was not easy. It posed enormous challenges to the people who embarked on it, selecting for those who could best survive under new conditions and who were the best at inventing technologies that would help their survival. At the end of this long selective process, the people who arrived in Australia were prepared to tackle one of the most difficult environments that our species has ever faced.

Humans were able to rise to this challenge. But the difficulty of getting to Australia and surviving once they arrived appears to have been too much for those more distant relatives of ours who also took the long trek from Africa almost 2 million years earlier. We will meet some of these earlier peoples, explore what happened to them, and see how our own species may have closed off evolutionary paths that might have been available to the earlier migrants, in our final chapter.

8

The San and the
Hobbits

Figure 122 A highly poisonous Komodo dragon, the world's largest monitor lizard, stalks across a hillside on the Indonesian island of Rinca. Larger versions of these dragons that roamed the nearby island of Flores were hunted by the tiny people known as Hobbits who lived there from at least 90,000 down to 11,000 years ago. Who were these people, how did they master such incredible challenges, and what might we have in common with them?

It is easy to die of thirst in the Kalahari Desert. Even the friendlier parts of this vast southern African desert are covered with thorny acacia scrub. Watercourses are rare and dry throughout much of the year. The land is flat and featureless, stretching out to the red and dusty horizon, so that it is extremely easy to lose one's way and wander helplessly in circles under the blazing African sun.

That is why I was glad I was in the company of a group of San women of the Nxaro tribe when we ventured out into the bush-covered red plains of the northern Kalahari.

In the last chapter we saw how modern humans fanned across Asia from Africa, becoming incredibly resourceful in the process. But the peoples they left behind in Africa were resourceful too. And, as we will see in this chapter, the remote relatives of ours that the people of the Great Migration encountered as they moved into Asia had their own resourcefulness. All of them shared properties of our multitalented and protean species. Here, in the book's final chapter, I will try to explore with you the essentials of the humanness that we all share and how that essential humanness has evolved. I will begin with the remarkable San, who never left Africa.

Hunters and gatherers

The San women, under the guidance of their leader, the tiny and cheerful grandmother Xhing Xhai, fanned out looking for things to eat. Most of the group were carrying babies, all were joking and gossiping. Between bouts of laughter they scrubbed at their teeth with twigs. Despite these distractions, within a few minutes they had begun to gather an astonishing amount of food.

Most of the women quickly became laden with a growing collection of herbs for cooking. Nxhwa found a collection of wild onions, lying just beneath an unremarkable patch of soil. Several of the group gathered around a thorn tree to dig up truffles, and quickly amassed a collection of these valuable fungi that would have fetched a king's ransom in Europe.

Figure 123 Women of the Nxaro tribe of San people in southern Botswana gather food from the bush while scrubbing their teeth with twigs.

Stefan, a Botswanan guide of European descent who was with us, had grown up with the San. The women badgered him for help because he was tall enough to be an expert flusher of jewel-beetles.

The local jewel beetles, of the family *Buprestidae,* are a vivid black and yellow. They live in the upper branches of thorn trees. The trick to catching them is to persuade Stefan to throw a large stick at the top of the tree. As the stick crashes into the branches the beetles fly away in all directions and everybody races after them to catch them in their hats.

Screams of laughter erupted as we scattered across the landscape. I had little luck, but within minutes the women had collected large numbers of the brilliant beetles. They crawled across their hands like animated sweets.

Other insects, such as clusters of tiny black grasshoppers that clung to the tree branches like grapes, were eagerly scooped up as contributions to the coming feast.

Gathering food in the bush is thirsty work. Xhing Xhai found a slender and unremarkable vine growing nearby. To my uneducated eye the vine was not obviously different from other small plants that formed the ground cover on this part of the bush. She recruited the others to help. Using their digging sticks they swiftly excavated a hole more than two feet deep. Xhing Xhai reached down into the dirt and pulled out a huge tuber. She and Stefan expertly scraped away the tuber's brown covering, revealing a snow-white interior. They carved the tuber up and passed pieces around.

The flesh of the tuber was cool, with an insubstantial texture. It was so full of liquid that eating it was like taking a drink of delicious, slightly radish-flavored water. As we exclaimed over this discovery, Xhing Xhai dug further down and pulled out another, deeper tuber from the same plant's root system. It was clear that we were in no danger of dying of thirst as long as we stayed in her company.

Back at the camp we ate slices of the truffles—in the process I dined on more truffles than I had consumed in my entire life up to that point. And the Buprestid beetles were roasted in an open fire and passed around to the eager members of the clan. The beetles' casings could easily be broken open to reveal their soft and runny greenish insides. I wish I could recommend them, but their bitter flavor and less than appetizing texture makes them a decidedly acquired taste. Nonetheless, they are rich in protein and fat, like the witchetty grubs eaten by Australian aborigines, and the San wolfed them down as eagerly as if they were chocolate-covered strawberries.

The San have adapted to this hunter-gatherer way of life because of brutal necessity. A thousand years ago Bantu tribes moved down into the ancestral home of the San in southern Africa, bringing agriculture and the ability to smelt iron for weapons and farm tools. They drove the San from the best hunting land and forced them to flee deep into the Kalahari. The small-bodied San, who had already been selected to survive on very little food if necessary, were able to draw on millennia of tradition to live and thrive on the unexpected bounty of the dry African bush.

Then the San were driven even further into the desert by the arrival of Europeans from the Netherlands in the sixteenth and seventeenth centuries. The Boers hunted the San like animals. As a result of this cruel invasion

Figure 124 A bumper crop of Buprestid jewel beetles. They are an acquired taste.

the San were reduced to shadowy, almost legendary inhabitants of areas so remote and undesirable that the invaders did not bother to follow them.

In the 1950s the South African war hero Laurens van der Post penetrated deep into the Kalahari on a BBC-funded expedition to film the San peoples. In his vivid *Lost World of the Kalahari* (1958) he recounts how vague stories of these tribes had peopled his childhood.[1] The San had been reduced to the status of half-understood legends by the cruelty and rapaciousness of van der Post's own ancestors. His accounts of his actual encounters with the San were embroidered with fancy, but he reintroduced the world to these remarkable people and helped to catalyze the establishment of the Central Kalahari Game Reserve. It is to be hoped that we denizens of our new and slightly more kindly world will be more likely to let the San survive than van der Post's ancestors did.

The emergence of the San from the mists of legend into the modern world has been difficult and painful. They have had decidedly mixed relations with the Bantu Tswana tribes that make up the majority of Botswana's population. The majority government has attempted to remove them forcibly from their ancestral home in the Central Kalahari Game Reserve. The government has also been clumsy and capricious in its enforcement of anti-poaching laws, targeting the San as they attempt to hunt in their ancestral lands.

Grim though this tale is, the San have survived much more in the course of the group's separate history, a period of at least 100,000 years that makes the history of Western civilization seem as trifling as a commercial break. They provide an essential glimpse into the whole sweep of our history as a species. Nowhere else in the world is there so vivid a demonstration of the genetic and social ties that bind all humans together over tens of thousands of years of time. To meet the San is to understand the essence of our shared humanity.

Figure 125 Stefan and Xhing Xhai sample the water-filled tuber she has found.

The mitochondrial "Eve"

Mitochondrial DNA studies show that the San form an important part of our family tree. The role that the San have played in our history became clear during the search for the mitochondrial Eve.

Just as it is possible to trace some Mongolian Y chromosomes back to a single ancestor who lived a thousand years ago, it is possible to trace the ancestry of all present-day human mitochondrial chromosomes back to one woman who lived in the distant past. Descendants of her mitochondrial chromosomes are carried by every human alive today. All the mitochondrial chromosomes that were carried by other women who lived at that time have been lost over the intervening millennia. All the differences among our present-day mitochondrial chromosomes have accumulated as a result of mutations since the time that ancestral woman lived; none date to before her time.

It is unfortunate that the woman in question has become known as the mitochondrial Eve. The world's press, and even some careless scientists and science writers, have repeatedly claimed that she was the ancestor of all present-day humans. This is not true, as other scientists have gone to great pains to point out with increasing levels of irritation. There were other women living at the time of the mitochondrial Eve, probably many of them in many different tribes, but by chance all the descendants of their mitochondrial chromosomes have become lost during the thousands of human generations from that time to this. At some point these other mitochondrial chromosomes all ended up in males or in females who had no children.

Even though the mitochondrial chromosomes of those other women have been lost by chance, some of the genetic information carried by their other chromosomes has survived. The women and the men of their tribes contributed some of their nuclear genes to people living today. Because of these contributions, our genetic legacy is far larger—and far more inclusive—than the little bit of mitochondrial DNA that was carried by the mitochondrial Eve.

Unfortunately, the name mitochondrial Eve has now become so firmly embedded, even in the scientific literature, that it is impossible to avoid using it. Let me emphasize again that she did exist, but that she was simply an average woman of her time and was not the mother of us all!

The best estimate of when the Eve lived is about 170,000 years ago, plus or minus 50,000 years or so. This happens to be roughly at the time that the fossil record suggests the first modern humans appeared in Africa.[2] And this coincidence has led to another widespread canard about the mitochondrial Eve, even more insidious than the one about how she was the ancestor of us all. This canard is that she was the first modern human.

Nonsense! Modern humans did not spring suddenly into being, like Athena from the brow of Zeus or Eve from Adam's rib. It is distressing to see how science writers (and once again some scientists) have carelessly applied the metaphors of the Book of Genesis to human origins.

There were far fewer people living at the time of the mitochondrial Eve than are alive today. The people of that time were all hunter-gatherers, like the present-day San. They survived from the animals they could catch, the tubers they could dig up, and the nuts, fruits, and wild honey they could gather. But they were not exactly like us. All of them were slightly different from any humans alive today. They were part of a complex and continuous evolutionary process that has given rise to the diversity of present-day humans and that will continue generating diversity and evolutionary change far into the future.

The mitochondrial Eve simply provides us with a kind of yardstick. She carried the common ancestor of all our present-day mitochondrial chromosomes, and we can estimate when and where she lived. But her existence tells us nothing about when the first modern humans appeared. That question is like asking when the first modern automobiles appeared.

Fossil evidence that modern humans lived in Africa approximately 160,000 years ago is dramatic and convincing. Modern-appearing skulls and other bones dating from that time were found by Tim White and his colleagues in Ethiopia.[3] Remarkably, the skulls had apparently been used for ritual purposes—the lower jaws had been removed and the brains had apparently been extracted. Stone tools and butchered animals bones were found along with the skulls. But White and his colleagues pointed out that these skulls were distinctly different from those of present-day humans—their brain cases were unusually large, and the skulls had slightly projecting brow ridges. They gave these fossils a new subspecies name, *idàltu*, which means "elder" in the local Afar language.

It is not surprising that *Homo sapiens idàltu* differed from us. A hundred and sixty thousand years is almost 10,000 human generations, plenty of time for genetic changes to accumulate. But let me again emphasize that even those ancient people are part of a continuum. Much older fossils of peoples who still show significant resemblances to modern humans have been found, especially in southern Africa.

Suppose we were to climb aboard a time machine and go back to the period of our mitochondrial Eve, roughly the time of *H. sapiens idàltu*, and suppose further that we could actually track the Eve down. She might have been a member of that Ethiopian tribe, but she was much more likely to have lived somewhere else in eastern or southern Africa.

If we could obtain mitochondrial DNA samples from her and from her contemporaries around the African continent, we would find that they differed among themselves because of accumulating mutations that had taken place in their past. The mitochondrial Eve's chromosome would simply be one of these sequences. Its only distinguishing characteristic was that it was the one that would survive and give rise to all of today's descendent mitochondrial chromosomes.

If we then traced this collection of lineages back in time, we would be able to infer the existence of an earlier mitochondrial Eve. Probably that Eve would have belonged to one of those earlier South African peoples that had some modern characteristics.

In short, there would have been nothing unusual about the woman who happened to carry the mitochondrial Eve chromosome. It is even unlikely that she lived at a particularly dramatic time in our evolutionary history. She was a tiny part of the great and continuing pageant of our evolution. The fact that she carried the mitochondrial chromosome that has been passed down to all of us is a statistical accident.

The San and the mitochondrial DNA tree

Even though we cannot use mitochondrial chromosomes to track down the mother of us all, because she never existed, they can be used quite unequivocally to trace the relationships among present-day groups of humans. The outlines of our genetic history were revealed when scientists built a vast tree of relatedness using a wide sampling of the mitochondrial chromosomes carried by present-day humans.

The shape of the tree is dictated by the mutations that have accumulated in our mitochondrial chromosomes as they have diverged from each other since the time of the Eve. The chromosomes at the tips of the deepest branches of this tree are those that have had the time to accumulate the greatest number of differences from the other branches. And, as Allan Wilson and his colleagues found out in the 1980s, all of the chromosomes on the deepest branches are found in African populations.

The branches that are deepest of all lead to the San and the central African pygmies. These isolated peoples have followed their own separate evolutionary trajectories from not long after the time of the mitochondrial Eve down to the present.

All other humans, even the Bantu invaders who drove the San from their ancestral homeland in the Karoo grasslands south of the Kalahari, are separated genetically from the San by a substantial evolutionary distance.

How long have the San followed their separate evolutionary path? If you believe the best estimate for the age of the mitochondrial Eve, and if you assume that the Eve lived at the time the San branched off from the rest of us, the answer would be approximately 170,000 years. But that answer is probably not right. Whenever we travel back in time using DNA as a yardstick, the uncertainty grows with each successive step. We must infer the pattern of branching further down the tree entirely from what we know about present-day sequences. The further back the branches are in time, the less certain we are about when they branched off and about the order of the branchings.

The uncertainty becomes greatest at the base of the tree—in this case, at the time when the mitochondrial Eve lived. It is possible that the San separated from the rest of us only 100,000 years ago instead of 170,000. At the other extreme, the ancestry of the San might have parted company from the rest of our species extremely far back in time, 250,000 years ago or more. Either of these scenarios is quite possible.

But, regardless of when the San branched away, it is clear that the common ancestor of the San and the rest of us was already *H. sapiens*. This is because the San are *H. sapiens* and so are the rest of us.

Is there something unique about the San, such as their amazing ability to find food in the unlikeliest places? Obviously not. Stefan comes from a family that had moved to Botswana from South Africa. He grew up with San children as playmates, learned two of their languages despite their intimidating glottal stops and clicks, and absorbed their bush lore from the time he was a toddler. He is also quite capable of surviving in the bush, and in fact he has hiked alone across the Kalahari and the Okavango delta for weeks at a time.

Could it be that there is something unique about the rest of us? Again, unlikely. The San languages are rich in words and concepts, their social relationships are complex, and they have proven perfectly able to learn about and take advantage of European technology.

The culture of the San

The latest eyeblink in the long history of the San has been enmeshed in politics and exploitation. These people of the Kalahari live in a diamond-rich area. The disastrous effort to resettle the San outside their desert home began with the desire of diamond companies to gain unrestricted access to the gems scattered throughout the desert. But tourism has proved to be a more stable source of revenue, and the Botswana government has halted diamond mining in the Central Kalahari Game Reserve. The San are starting to move back to their homes, and there are now attempts to re-establish some approximation of their original bushland ecosystem.

Figure 126 This fence divides the ecological reserve where the Nxaro live from a hunting preserve run by Europeans.

But Botswana is an enormous place, and many conflicting trends are taking place simultaneously. The Nxaro group that I visited live outside the boundaries of the Central Kalahari Reserve, in a region that is a patchwork of farming, ecotourism, and private game reserves.

White rhinos have recently been reintroduced into the area, and other large animals are starting to become more common. But the artificiality of the whole jury-rigged arrangement is striking. At the same time as we were photographing antelope and giraffes near the Nxaro village, a few kilometers away the same species were being shot by European hunters at a cost of thousands of Euros for each trophy.

Because of these recent changes, only fragments are left of the unimaginably ancient San hunting tradition. On the evening before our foray into the bush, we had watched as the Nxaro tribespeople gathered around a fire and danced a re-enactment of an ancient hunt. A young boy, wearing anklets made of dozens of white moth cocoons that rattled in rhythm with the drums, positioned two long poles on his head to imitate the long straight horns of the gemsbok. Two older men stalked him and mimed a kill, while the onlookers shouted with excitement.

Alas, there are no gemsbok in the area now, though they are scheduled to be reintroduced. The Nxaro dance was a window into a world that no longer exists. And the San, closed off by their difficult languages from the rest of Botswanan society, may soon be relegated to life as an ecotourism curiosity, condemned to act out shadow plays about the oldest continuous human culture on the planet.

The world that the San knew is preserved in their rock paintings. It is hard to date such paintings, but some unequivocal ages of more than 3,000 years have been obtained. These paintings predate the time when the San were pushed into the Kalahari by the arrival of the Bantu. The half-human, half-

Figure 127 A young Nxaro boy mimics a gemsbok in a ritual hunting dance.

animal figures in some of the paintings may be direct evidence of the antiquity of the dances that I saw, in which people take on the properties of animals. Or they may represent animist religious traditions that have now largely been lost.

The oldest San paintings that can be dated with certainty are in the Drakensberg Range in eastern South Africa, but paint chips buried in the soil near the paintings may be centuries older. The paintings from which these buried chips come, along with unknown numbers of generations of earlier paintings, have been destroyed by weathering.

There are also tantalizing hints of an astonishingly ancient San artistic tradition. A rock with bits of ocher paint adhering to it has been found in a cave near South Africa's coast. The soil layer in which the rock was embedded comes from a stratum dated at more than 20,000 years before the present. And, in a cave in Namibia near Africa's dry southwest coast, German archeologists have found fragments of paintings that were first tentatively dated at 28,000 years ago. It now appears that the paintings may be much more recent.[4]

Even if these oldest dates are wrong, it is clear that the San have an ancient artistic tradition. Their efforts to depict the world around them certainly precede any sustained contacts with cultures from the north. The San have also developed sophisticated hunting weapons and decorative arts. In general, their cultural developments show remarkable parallels with those that were taking place further to the north in Africa and in Europe.

Parallel cultures and the overthrow of Eurocentrism

The parallel development of cultures is a theme that emerges repeatedly as we examine the history of our species. Paleoanthropologists study human prehistory through the fossil record and cultural artifacts. The founders of this field, in the late nineteenth and early twentieth centuries, concentrated their efforts on Europe—not surprisingly, since they happened to be Europeans and they were surrounded by conveniently situated European archeological sites. Much effort was put into investigating the cultures of the

Neanderthals, the big-brained and large-boned people who lived in northern Europe during the recent ice ages, and those of the more delicate-boned Cro-Magnon people who replaced them. The Cro-Magnons were considered by these paleoanthropologists to be the first truly modern humans. This conclusion was reinforced by the discovery, starting in 1870, of many magnificent Cro-Magnon cave paintings.

It seemed perfectly obvious to these scientists of the nineteenth century that much of the important cultural development of our species had taken place in Europe. It was therefore quite easy for them to conclude that modern Europeans are the most highly evolved and the most culturally advanced peoples on the planet. As the inheritors of the noble Cro-Magnon tradition, it seemed clear to them that modern Europeans richly deserve the right to fan out across the world and, as Kipling memorably expressed it, exert their dominance over "lesser breeds without the law."

But, starting in the late nineteenth century, a series of disorienting discoveries began to put human history into a much larger context. These discoveries eventually destroyed the Eurocentric view.

Darwin had noted that our closest primate relatives live in Africa. Because closely related groups of animals tend to live near each other, he concluded it was likely that modern humans, too, first appeared in Africa.

Darwin's prediction was ignored by anthropologists of the time. The European Neanderthals were assumed to be the progenitors of modern humans, who must therefore have evolved in Europe. But in 1891 the Dutch doctor Eugene Dubois found a skullcap of a small-brained human progenitor embedded in a muddy bank of the Solo River in central Java. Many more remains of this progenitor, later called *Homo erectus*, were found in Java, in several parts of China, in eastern Africa, and most recently in the western Asian country of Georgia.

In agreement with Darwin's prediction, *H. erectus* originated in Africa. We now know that these intelligent prehumans migrated into Asia out of Africa roughly 2 million years ago. Unlike the later Great Migrants, they managed to cross the desert barriers of the Middle East during periods when it was more like the East African savannah. And we now know that the later Neanderthals of Europe also originated in Africa, probably from a *H. erectus* lineage. It is likely that the ancestors of the Neanderthals made their way into Europe and the Middle East starting about a million years ago.

African human origins were soon driven much further back in time by a find in South Africa. In 1924 Raymond Dart, an Australian-born anatomist, was sent a scientific oddity by the owner of a nearby quarry. It was a tiny ape-like skull embedded in a chunk of quarry stone. Dart thought at first that the skull was that of a fossil ape, but when he finally freed the lower jaw from the rest of the skull he found that its teeth were not apelike. He had expected to see a typical ape dentition, with prominent canines and with cusps on the molars and premolars that lock the upper and lower jaws together. Instead he found that, like the teeth of modern humans, the teeth of this "southern ape" *Australopithecus* had small canines and flattened molars. This meant that its lower jaw could have shifted from side to side in order to grind food rather than just crush it.

Dart's skull was probably less than 2 million years old, but since its discovery a variety of Australopithecines have been found that date back more than 4 million years. This diverse set of African species has added to the growing number of species in our own ancestry, as distinct from the ancestry of the great apes such as the chimpanzees and gorillas. Together, we and our ancestors, along with other branches on the human-like family tree such as the Australopithecines, are now put in the subtribe Hominina and called hominans. All the hominans and chimpanzees grouped together make up the tribe Hominini and are called hominins.

After all these discoveries, and after the discovery of precise ways to date the fossil finds in their context, it is clear that both the fossil and the molecular evidence are unequivocal about our African origin. Not only did the many-branched lineages that led to the Australopithecines and to *H. erectus* arise in Africa, but so did the lineage that led to modern humans.

By the early twentieth century it had become clear that Darwin was right about our origins, as he was about so many other things. The Eurocentrics had been handed a decisive defeat.

The great leap forward—not!

But what about the origins of advanced human culture? These same Euro-centric paleoanthropologists had carefully documented a burst of cultural advances that took place at the same time as the first modern humans arrived in western Europe, about 30,000 years ago. These dramatic advances have been called the "great leap forward" by University of California, Los Angeles, physiologist and ecologist Jared Diamond.

Sally McBrearty and Alison Brooks of the University of Connecticut and George Washington University emphatically disagree with the idea of a European great leap forward. They have used the growing archeological record of sub-Saharan Africa to trace the real, and much older, origins of these cultural inventions.[5]

It is true that there is nothing in sub-Saharan Africa that corresponds to the detailed and expressive cave paintings of Altamira in Spain and of Lascaux and Chauvet in France. But the Cro-Magnon paintings are primarily of animals, many of them now extinct in Europe. There are almost no human figures among those great herds of gracefully rendered Ice Age animals.

In contrast, even the oldest San rock paintings show many representations of human dancers and hunters. By this measure their art is much more oriented toward their cultural lives than the pictures drawn by the Cro-Magnons. Of course, most of the surviving examples of these San paintings are far more recent than the European cave paintings. If we discount the doubtful evidence from the Namibian cave, we simply do not know how long ago the ancestors of the San began to paint this representational art on the rock walls of the Karoo and the Kalahari. Did the San learn the idea of representing the human form from other tribes, or did they invent the idea independently? Again, we cannot answer that question.

Thus, the true origins of representational art, and the number of times that the idea of drawing people and animals occurred independently to various human groups, remain a great puzzle. But haunting suggestions of an ancient origin of representative art continue to turn up.

Figure 128 Four views of the Berekhat Ram figurine, possibly the oldest representation of a human form yet discovered. The scale represents 1 centimeter. (From Francesco d'Errico and April Nowell, "A New Look at the Berekhat Ram Figurine, Implications for the Origins of Symbolism," *Cambridge Archaeological Journal* 10 (2000): 123-67, Figure 1.) Courtesy of Francesco d'Errico/*Cambridge Archaeological Journal.*

Consider the "Venus of Berekhat Ram" that was discovered in 1981 at a dig in northern Israel. This little stone figurine looks like a cruder version of the voluptuous female figures that have been found from late Stone Age sites at various places around the Mediterranean. These later venuses, however, are a few thousand years old, while the Berekhat Ram figure was found sandwiched between two layers of volcanic ash that clearly assign it an age of a quarter of a million years.

Microscopic and electron microscope examination showed that the Berekhat Ram figure had indeed been shaped and incised by tools in several places, though the authors are careful to point out that it may not actually be a figurine. Was this ancient Venus really a representation of a zaftig female form? Or was somebody, presumably a *H. erectus*, simply noodling around with a piece of rock? A definitive answer awaits similar discoveries.

It is clear from the surveys and discoveries of McBrearty and Brooks that many important cultural advances really did not originate in Europe. Knife-like stone blades, obviously the work of humans, have been found at sites in several parts of Africa that go back at least 300,000 years. These blades are too small and narrow to be held comfortably in the hand. Rows of the blades were probably embedded in wood using pitch or resin, to produce a sharp multi-bladed cutting tool. The earliest stone pestles for grinding food also date to this time, and these pestles are often associated with worn pieces of ocher that probably served as sources of pigment.

Arrowheads a quarter of a million years old have been found in West Africa, predating the first undoubted modern human skeletal remains. Barbed fish hooks were invented roughly 100,000 years ago. A piece of ocher inscribed with repeated patterns of squares and diamonds was found recently in Blombos Cave in South Africa, and has been dated to 75,000 years ago. Did this incised object represent an early attempt at decorative art, was it used to mark the passage of time, or did it have some entirely different purpose?

European culture before modern humans

McBrearty and Brooks concentrated on the previously neglected story of sub-Saharan Africa. But there are hints that Africa was not the only place where important cultural advances were being made. Some of those advances even took place long before the emergence of *H. sapiens*.

One of the hominan species that have recently been shown to possess advanced culture was the shadowy group of people who preceded the Neanderthals in Europe.

The history of hominans in Europe is long and complicated. The oldest European human bone fossil, from a cave near Atapuerca in northern Spain, is a fragment of lower jaw associated with stone tools. The fragment has been dated to a remarkable 1.1 to 1.2 million years ago.[6]

Thus, more than a million years ago the first hominans arrived in the great European peninsula from Africa and the Middle East. They spread rapidly across southern Europe, from Greece to Spain. Much of the rest of Europe was eventually colonized by a diverse collection of these peoples.

It is clear that these early Europeans were not our direct ancestors. Because they tended to have some Neanderthal-like features, they may have been the ancestors of the Neanderthals of Europe and the Middle East. The Neanderthals themselves lived much closer to our own time. The earliest skeletal remains in Europe that can be unequivocally classified as Neanderthal date to about

130,000 years ago. Neanderthals spread subsequently and were enormously successful—their remains have been found in Gibraltar, in northern and eastern Europe, and as far east as Iran. The much older peoples whom they seem to have displaced are now collectively referred to as pre-Neanderthals.

Then, starting about 30,000 years ago, the European Neanderthals were displaced in their turn by the latest wave of immigrants to arrive from Africa via the Middle East. These immigrants included the Cro-Magnon people and other modern humans.

This unfolding scenario from the fossil and archeological record of Europe dovetails neatly with the emerging molecular record. The only ancient hominan bones from which meaningful amounts of DNA have been extracted so far are Neanderthal and Cro-Magnon remains from Europe. The DNA was preserved because these bones stayed relatively cool and dry. The growing number of Neanderthal DNA sequences now becoming available place the Neanderthals clearly in a separate lineage from our own, while the Cro-Magnons are definitely modern humans with some affinities to present-day Middle-Eastern populations.

The DNA branch point that divides the Neanderthal and modern human lineages can be dated to about 600,000 years ago. But the errors on this number are so large that the time at which the Neanderthals branched off from us could be as much as a million years ago, roughly the same time as the appearance of the earliest European hominan fossils. Or it may be that there were successive waves of hominans entering Europe that eventually recombined and reassorted with earlier migrants to become the Neanderthals. There is fascinating evidence that humans and chimpanzees did not speciate instantaneously 6 million years ago, but that the speciation process was much more complicated and lengthy. Perhaps the origin of the Neanderthals will be discovered to be equally complex.[7]

The Neanderthals were remarkably successful* and mobile. They made decorative objects, and the Neanderthal tribes in the interior of France acquired,

---

* The recent astonishing discovery of an ancient mitochondrial sequence, more than twice as divergent from ours as the Neanderthal sequences, from a single finger bone found in a Siberian cave that has signs of Neanderthal occupancy, may give us a first glimpse of the relationship between the Neanderthals and the pre-Neanderthals. Stay tuned!

perhaps through trade, objects such as sea shells that they used to make neck-laces and other adornments. Neanderthals living in the area that would eventually become northern Iran buried heaps of flowers along with the bodies of their dead. In Spain, long before the arrival of modern humans, Neanderthals used sea shells to mix pigments, perhaps for decoration. But there is no evidence as yet that the Neanderthals carved any representations of human figures, or drew and painted on the walls of caves. These skills were introduced into Europe by the modern humans who displaced the Neanderthals.

The argument has been made repeatedly that the Neanderthals were simply not clever enough to have made such sophisticated objects and paintings on their own. Perhaps, it has been suggested, they obtained the objects that are associated with them from nearby modern human tribes. Alternatively, the modern humans might have driven the Neanderthals out of the caves, so that the artifacts actually belonged to the modern humans that displaced them. Or perhaps the Neanderthals sneaked over to nearby caves and pilfered the objects from modern humans. Arguing against all these rather unkind scenarios is the fact that the artifacts associated with Neanderthal remains are distinctly different from those made by modern humans that have been dated to approximately the same time.

The most powerful argument that the Neanderthals were perfectly capable of making their sophisticated tools and decorative objects comes from a glimpse of the capabilities of their remote pre-Neanderthal predecessors. Refreshingly, this glimpse involves artifacts that are different from those boring-looking stone tools that archeologists spend so much time exhuming and arguing over.

It is not surprising that most of the tools and artifacts found by pale-ontologists are made of stone or shell, because wooden and other organic artifacts soon rot or weather away. Occasionally, however, some of these organic materials are accidentally preserved. Over the past century a few such fragile artifacts belonging to pre-Neanderthals have been found in Europe, including a possible wooden spear tip in England and what looks very much like a complete but poorly preserved spear in Germany. But these fragmentary discoveries were dismissed by authorities as likely to be nothing but primitive digging sticks.

Then, in 1997, Hartmut Thieme of the Institute for the Preservation of Historical Objects in Hanover reported an astonishing discovery in a gigantic open-pit peat mine in central Germany. Peat, which is euphemistically known as brown coal, is actually plant material that is in the process of becoming coal. It is dug in vast amounts at many places around the world, leaving behind immense open pits that pock the landscape like some skin disease that will take centuries to heal.

Thieme and his group had carefully monitored this particular pit mine, where gigantic rotary shovels dig out the peat around the clock. They and others had already found bits of wooden artifacts dating to half a million years ago, but the nature of the artifacts was unclear and their context was lacking. Then the shovels uncovered an artifact-rich area that had obviously been lived in by hunters.

Amidst thousands of bones of ancient horses, many of which showed signs of butchering, Thieme and his co-workers found three 2-meter-long throwing spears that could be dated to 400,000 years ago.[8] The spears were preserved because the slow decay of the peat had kept oxygen away, and had prevented the wood from being broken down.

Made from young spruce trees, the spears were highly sophisticated in their design. Like modern javelins their centers of balance were about a third of the way back from their points. And the points themselves were carved from the bases of the trees. As modern Stone Age hunters know, the base of a sapling has the hardest wood and makes the most effective point.

The San of the Kalahari and many other present-day hunter-gatherer groups make similar spears. The San spears are only a little over a meter long, because they are small people, but they can use them to bring down large animals like gemsbok.

Thieme's accidental discovery makes one wonder how many other essential glimpses of ancient Europe have been lost from peat mines. What treasures have been crunched up by the rotary shovels as they gobble up the landscape to feed some of the world's most polluting power plants with a fuel that has one fifth of the energy per ton as hard coal?

Luckily, the spears that Thieme and his crew did manage to rescue provide clear evidence that almost half a million years ago the pre-Neanderthals of northern Europe were hunting at a level of sophistication equivalent to that seen among the San hunters of the present time. The pre-Neanderthals must have been able to conmunicate with each other in the same way as present-day hunters do—it is otherwise difficult to imagine how they could have killed big fast animals like horses, butchered them, and carried the pieces of their kills substantial distances to a central gathering place.

The myth of human uniqueness

Thieme's find suggests that some of our relatives were cleverer than we give them credit for, even though they may have diverged from us more than a million years ago. It is further evidence against the deep-rooted notion that modern humans have somehow taken a special evolutionary road that sets us apart from other living things.

Darwin firmly rejected the view that humans are special and outside the laws that govern evolution. He insisted that humans have been shaped by the same evolutionary processes that have acted on all other organisms. He emphasized that natural selection, especially sexual selection, must have acted on heritable variation that was present in populations of our ancestors. Even though we are clearly different from other animals, we evolved by quite conventional means.

Darwin's contemporary Alfred Russel Wallace had independently reached conclusions similar to Darwin's about the ability of natural selection to shape the natural world. The idea came to him in 1858 during a bout of fever, when he was on the island of Ternate in the East Indies. After he recovered he hastily dashed off a short paper, and mailed it to Darwin for criticism. It was the arrival of Wallace's manuscript that finally jolted Darwin into protecting his priority by publishing an abstract of his long-projected multivolume *magnum opus* about natural selection. The abstract was the *Origin of Species*.

But Wallace, unlike the pragmatic Darwin, had an intense interest in spirituality and the nature of the human soul. He took a view different from Darwin's about the evolutionary story of our species, insisting that there had to be something unique about our evolution.[9]

Wallace was led to this conclusion by a growing interest in spiritualism, sparked by observations of the religious practices of peoples he had visited in his travels in South America and Malaya. He was also most impressed by the results of some experiments he carried out with hypnosis. He concluded that there were powers of the mind that could not be explained by the ordinary processes of evolution. Human evolution must, he felt strongly, have some kind of a supernatural dimension.

Wallace's views were in their essence similar to those of the Catholic Church, which since the time of Pope Pius XII has been clear in its statements that humans are distinguishable from other animals by the possession of a soul and of the ability to believe in and accept a higher power. More recently the Church (or at least the Vatican) has specifically accepted the scientific mechanisms of evolution, but always with the caveat that the creation of humans constitutes a special case because of our relationship to a higher power.

The Church is of course confident that the higher power involved is God, while Wallace was far less certain of the nature of the spiritual dimension of our species. Wallace seems to have believed in some kind of supernatural evolutionary continuity of lesser deities that lie between God and man. He suggested that some of these lesser deities have assisted us on our evolutionary way:

An eminent French critic, M. Claparède, [suggests that I] continually call in the aid of "une Force supérieure," the capital F, meaning I imagine that this "higher Force" is the Deity. I can only explain this misconception by the incapacity of the modern cultivated mind to realise the existence of any higher intelligence between itself and Deity. Angels and archangels, spirits and demons, have been so long banished from our belief as to have become actually unthinkable as actual existences, and nothing in modern philosophy takes their place. Yet the grand law of "continuity," the last outcome of modern science, which seems absolute throughout the realms of matter, force, and mind, so far as we can explore them, cannot surely fail to be true beyond the narrow sphere of our vision, [not leaving] an infinite chasm between man and the Great Mind of the universe. Such a supposition seems to me in the highest degree improbable.

Darwin took Wallace to task for these and similarly vague and goofy ideas about human evolution. Indeed, Wallace's scientifically unorthodox views hindered his own scientific career and help to explain why he has always played second fiddle to Darwin.

The myth of human uniqueness, that we are a special case, was only slightly dented by the discovery of the Neanderthal culture, since it was possible that the Neanderthals had simply copied from modern humans. It was decidedly pummeled by Thieme's discovery that the pre-Neanderthals, who certainly had no contact with modern humans, were expert hunters. But it has been smashed to pieces by the discovery of the Hobbits, an advanced people who may have parted company with our own immediate ancestors and begun to pursue an independent evolutionary path more than 2 million years ago. A separate Hobbit lineage may be traceable more than twice as far back as that of the Neanderthals, and yet the Hobbits too had attributes that we think of as uniquely human.

Here there be dragons

The Hobbits, perhaps our most astonishing relatives, lived on the island of Flores in Indonesia. Let me begin their story by setting the scene.

As we saw in the first chapter, Flores is one of the Lesser Sunda Islands, a chain of islands that stretches from Java to the western end of New Guinea. Each of the islands in the chain has its own distinct character. Lombok and Sumbawa are large islands with a mix of open forest and grassland. Sumbawa is dominated by the great cone of Tambora, the volcano that produced the "year without a summer" after its vast eruption in 1812. Komodo and Rinca are smaller, with patches of gallery forest and much open grassland. They are the last refuge of the Komodo dragon, the world's largest lizard.

The island of Rinca is especially gorgeous, and is as full of dragons as any medieval map-maker could wish. When I recently stepped ashore at Rinca, which now forms part of Komodo National Park, I was immediately confronted by a dragon that forced me to leap to safety on a convenient ledge. The dragon then crawled out onto the dock, producing an indescribable silky sibilant clicking sound as it slithered over the boards. It sent the boat crews scurrying for shelter before it turned and moved regally back to its island.

Later, as I hiked Rinca's grassy uplands, I continued to be a bit nervous, since the dragons have been known to eat villagers and the occasional tourist. But the thriving dragon population has adjusted to tourism, and the guides have learned that the lizards can be pushed around with forked sticks, like living pucks in a game of shuffleboard.

The dragons are relict populations of much more numerous (and much larger) dragons in the past that used to swarm over the Lesser Sundas east of Wallace's Line. The first hominans who ventured along this island chain found themselves up to their necks in dragons.

Flores, to the east of Rinca, is a large and mountainous island with more abundant rainfall and thriving crops of rice, spices, coffee, and tea. It is dominated by several active volcanoes. In 1992 a severe earthquake and tidal wave destroyed the village of Maumere on the island's north coast, killing almost 2,000 people.

You will recall from Chapter 3 that the animals and plants that live on the Lesser Sunda Islands have been shaped by the collision of two great provinces of the living world. The most obvious result of this collision has been Wallace's Line, 500 kilometers west of Flores, which separates the Southeast Asian from the Australasian flora and fauna.

The humans who currently live on the Lesser Sunda chain have also been influenced by the collision of geography and history. The island of Bali has retained its early Hindu culture, because of its success at repelling invaders. The islands from Lombok to Sumbawa are all converts to Islam, because of the influence of traders and conquerors from Mogul India. Flores is largely Christian, influenced by the early arrival of Catholic missionaries from Portugal, though western Flores is now becoming more Muslim because of immigration from elsewhere in Indonesia.

These different populations show dramatic cultural differences. Bali is famous for its charming Hindu festivals and for ceremonies interwoven with everyday life. On Sumbawa life is a great deal less happy-go-lucky. A local

Figure 129 (*overleaf*) Approximately a million years ago, when the first hominans arrived in this part of the world, these mangrove-rimmed islands that lie to the west of Flores in Indonesia were linked together by dry land bridges.

tradition of female genital mutilation ("circumcision") persists on the island. This repellent ritual appears to predate the Muslim conversion. On nearby Christian Flores female circumcision is no longer permitted.

Today's cultural differences cause dramatic disconnects from one island to the next, but this cultural diversity is a pale reflection of the differences that separated the islands' hominan populations in the distant past. As we saw in the previous chapter those past diversities were generated, not by the invasion of different religions, but by the migratory movements of hominans and later of modern humans along the length of the island chain.

Modern humans arrived on the Sunda Shelf more than 40,000 years ago. Their arrival was followed by the disappearance of the *H. erectus* tribes that had lived there for a million and a half years. There are striking parallels between these events and the arrival in Europe of the Cro-Magnons and other modern groups, which took place at around the same time. The Cro-Magnon arrival in Europe was followed, somewhere between 2,000 and 10,000 years later, by the disappearance of the Neanderthals. It seems that a similar sequence of events may have occurred in Southeast Asia. Did the invading groups of modern humans drive the last *H. erectus* to extinction? If so, then everywhere they went modern humans seem to have been extremely effective, and merciless, competitors!

The clever little Hobbits

The complex prehistory of humans and other hominans in Southeast Asia has been made even more fascinating by the discovery of the Hobbits on the island of Flores.

The first hint that any sort of hominan had penetrated to this remote island came in 1970, with the publication of a report of stone tools at a site called Mata Menge in central Flores. The site, part of the Soa volcanic basin, is an open area

Figure 130 (*opposite*) This Komodo dragon, *Varanus komodoensis*, stalks out onto the dock at the island of Rinca, sending the boat crews scurrying. The first hominans to arrive on Flores found even larger relatives of today's Komodo dragons awaiting them. They were nonetheless able to exploit these terrifying and poisonous animals for food.

carved by shallow gullies cut through alternating layers of sandstone and volcanic deposits. The tools were found in sandstone strata near bones of extinct elephant-like animals, pygmy *Stegodon* that were about the size of a cow.

The first discoverers of the *Stegodon* bones were Soa tribespeople, who traditionally burn off the area every March during hunts. In 1953 they collected Stegodon bones that had been washed out of the deposits and presented them to the local Raja of Nagekeo. The Raja in turn alerted Theodor Verhoeven, a Catholic priest and amateur archaeologist.

Verhoeven carried out excavations at Mata Menge in 1963. He discovered the crude stone tools among the plentiful *Stegodon* bones, and attributed them to *H. erectus* because they resembled tools that had been found associated with *H. erectus* bones on Java. But he was unable to date the tools convincingly, and his papers about them were ignored.

Figure 131 The Ngada tribal village of Luba on Flores, more than 300 years old, is dominated by the mildly active volcano Gunung Inerie. The shapes of the houses reflect the shapes of the dozens of volcanoes and cinder cones that dominate the lives of the people of Flores in this geologically active part of the world.

Figure 132   An effigy on the roof of a house in the nearby village of Bena signals the importance of its owner.

Verhoeven died in 1990, and it was not until 1998 that other scientists finally followed up on his observations. A group led by Michael Morwood of Australia's University of New England and Fachroel Aziz from Bandung on Java were able to date the volcanic layers above and below the stone tools at Mata Menge.[10] These and other dating methods yielded an astonishingly old date of approximately 800,000 years. Similar stone tools, some as old as 1.1 million years, were found later at a number of nearby sites.

Morwood and his colleagues agreed with Verhoeven's conclusion that the makers of these stone tools must have been *H. erectus*. This was because the only hominans known to be living in the Sunda islands at this early date were *H. erectus*, and the *H. erectus* of Java had used similar tools. But even so their conclusion was startling, because deep-water channels separate Flores from the islands to the west. Even during the severest Ice Age the gaps between the islands would never have closed. The only way that these hominans could have crossed was by boat or—far less likely—by clinging to trees uprooted

Figure 133 Children on the Muslim island of Sumbawa in Indonesia. Nearby islands in the same chain are Hindu or Christian.

by a storm. And, because of the strong currents that raced through the channels, they would probably have had to cross between islands further north and work their way south to Flores.

Regardless of how the migrating hominans managed to accomplish the feat, it was clear that they had somehow made their way to Flores hundreds of thousands of years before the arrival of modern humans. But there were no skeletal remains to give us clues about what they were actually like.

Nothing further was found out about the early inhabitants of Flores until 2003. And then Morwood and his Australian–Indonesian team made one of the most breathtaking discoveries in all human prehistory.

They were carrying out deep and technically challenging excavations in one of the great open-mouthed limestone caves in the mountainous central part of the island. The cave, called Liang Bua ("cool cave" in the local language), is similar in its geological origin to the Niah caves of Borneo. Such large limestone caves, first carved out by groundwater seeps and then exposed to the outside by subsequent erosion, are common throughout Southeast Asia.

About half a million years ago the interior of Liang Bua Cave was exposed to the outside world for the first time when a nearby river's path shifted so that it flowed past the mountain containing the cave. The river broke open part of the cave's thin limestone shell, illuminating the clusters of ancient stalactites on the cave's roof for the first time.

In the hundreds of thousands of years since that time the river has continued to sculpt the landscape. Currently it flows through a nearby valley about 20 meters below the level of the cave. And because the cave is opened up to the outside, it has been repeatedly flooded by the torrential rains of the region. The result of all this flooding has been an accumulation of great quantities of wet silt on the cave's floor. For much of the time during which the cave has been open to the elements there has been at least one deep pool of water on the floor.

Back in the 1960s and 1970s Father Verhoeven and Indonesian excavators had dug a little way down into the cave floor deposits. They found remains of modern humans near the surface, along with bones of the animals that they had killed. But they were stopped by a layer of hard volcanic tufa that seemed to mark the base of the deposits.

Morwood's team was able to break through this thick layer, finding immense deposits of silt below. But digging down further proved to be technically difficult. At every stage in the excavations the pits had to be shored up by massive wooden scaffolding, and even so there was one sudden collapse that nearly caught some of the workers.

As Morwood's team dug further down through the wet layers they began to find the bones of animals, some now extinct on Flores, that also showed signs of having been butchered. Along with the bones they found stone tools that could have been used to dismember the animals. These tools resembled the very old tools that had been found earlier at Mata Menge and other nearby non-cave sites.

Early in these new excavations they unearthed a tiny hominan radius, the small bone of a lower arm. The radius was extremely fragile, making it a challenge to free it from the muck that surrounded it. It was tiny but proportionately unusually thick, puzzlingly different from the equivalent bone of a modern human. This exciting discovery spurred them to continue digging down through the mud.

The animal remains that they found during this deep excavation conjured up a world very different from present-day Flores. There were plentiful bones of the extinct elephant relatives known as Stegodon. These animals occupied a distinct branch of the elephant family tree, different from both the extinct mammoths and the living African and Asian elephants. Stegodons were plentiful throughout Southeast Asia, and indeed some of them were mighty-tusked beasts that ranked among the largest of the elephant family. The Stegodons of Flores, however, were small. The ones that might have been killed by *H. erectus* at Mata Menge were true pygmy Stegodon, but by the time that the Liang Bua deposits had been laid down these extremely small elephants had evolved into a slightly larger subspecies.

Of course, the word small when applied to elephants is a decidedly relative term—even the smallest of these animals were as large as a water buffalo, and they were still equipped with long wicked tusks for defense. Despite their potential fierceness, cut-marks on the bones in the cave deposits clearly

Figure 134 The Soa Basin in central Flores at dawn. In the background to the right is the active Abulobo volcano, one of a large number of active and extinct volcanoes that ring the area.

Figure 135 Abulobo, overlooking the Soa Basin, vents steam in the early morning light.

showed that the animals had been disarticulated and butchered by someone. And it was striking that most of the skeletons that the team of diggers encountered belonged to young animals. It seemed likely that these Stegodon babies had been easier for their unknown hunters to kill than the more dangerous adults.

As the excavators dug down, the bones of even more daunting creatures, supersized Komodo dragons, began to turn up. The present-day Komodo dragons, 3 meters long and weighing as much as 140 kilograms, are determined predators. The can attack and kill water buffalo. Their bite poisons their prey with secretions from glands in their jaws that cause severe bleeding. The secretions also lull the prey into calmly accepting their imminent doom.

During the time that the excavators were exploring, the ancestors of the Komodo dragons were widespread throughout Australasia. And some of those extinct ancestors were much larger than the present-day dragons.

309

Figure 136 Father Theodor Verhoeven's dig at Mata Menge, where he found stone tools mixed with Stegodon bones. Verhoeven hypothesized that early hominans, probably Homo erectus allied to Java Man, had lived in the area, even though these people would have had to cross deep-water passages in order to reach Flores.

The giant Australian *Megalania* went (or was driven) extinct at the same time as modern humans arrived in Australia. It was dinosaur-sized, probably 6 or more meters in length. The Komodo dragon bones found in the Liang Bua cave belonged to animals not quite as large but still bigger than those on Komodo and Rinca today. The hunters of these animals must have been expert, fearless, and well organized.

Scattered among all these bones were more crude stone tools, like those that had been found at Mata Menge. And the people who had killed the animals were quite capable of cooking their kills. Fire-charred stones, including the remnants of a distinct fire ring, were found along with the bones.

In September 2003 the Australian members of the team had left for the season. The Indonesian members, led by Thomas Sutikna, continued the excavation. Then, just weeks after the Australians had left, the Indonesians discovered the remains of an almost complete hominan skeleton.[11]

The team excavated the bones with exquisite care, first digging out the entire mass of deposits that contained it and transporting it to a nearby hotel. Prolonged soaking in Liang Bua's wet environment had turned the bones, in the memorable phrase of one of the expedition's members, to a kind of "mashed potatoes with the consistency of wet blotting paper." In order to harden these exquisitely valuable bits of mush, each of them had to be coated with glue diluted with acetone.

The little female skeleton was fairly complete, some of its bones still joined together by bits of cartilage. Apparently she had fallen into one of the deep pools in the cave and been preserved from decay in the oxygen-free

Figure 137   This is how the Liang Bua excavation appeared in early September 2009, looking out through the cave's opening to the river valley beyond. The shaft where the complete Hobbit skeleton was recovered is on the far right. Earth from the excavations, carefully sieved to recover thousands of tiny animal bones, is piled up to the left, awaiting the time when it can be used to fill up the excavations again.

waters at the pool's bottom. And it was clear that, tiny as she was, she was a mature adult—her wisdom teeth had all erupted.

Pieces from at least eight other skeletons have since been found in the same shaft. None of these people were more than a meter tall, which made them much shorter than present-day African pygmies. The Smithsonian's Matt Tocheri, whom I met at Liang Bua, has found that their wristbones most closely resemble those of apes and the tiny Australopithecines of Africa, which had less freedom of wrist movement than modern humans.[12] The Hobbits' arms were extremely long relative to their tiny bodies, and their arm and leg bones were proportionately thick, with dimensions that lay halfway between those of *H. erectus* and the sturdy little African Australopithecines.

The skull from the complete skeleton is marked by prominent brow ridges, giving it an appearance somewhat like the skulls of *H. erectus*, but with additional odd features that are once more reminiscent of the Australopithecines. Like most other hominans, these little people were fully upright. Signals of upright posture are given by the angles of the skull and jaw, and the position of the hole in the bottom of the skull through which the spinal cord passes. These are almost indistinguishable from those of modern humans.

Much of our brain-power resides in our large cerebral cortex, the layer of gray matter that covers our cerebral hemispheres. Our cortex is convoluted and fissured, giving it a greater area than if it were a simple smooth coating. The Hobbits' brains were a third the volume of those of modern humans, even smaller than adult chimpanzees' brains.[13] But their brains seem to have packed far more intellectual punch than the brains of chimpanzees. A CAT scanner was used to visualize the impressions on the inside of the Hobbit's skull that had been left behind by her long-vanished brain. Judging by their complexity, her cortex was even more deeply fissured and wrinkled than a modern human cortex. Although the Hobbits' brains were tiny, they were no intellectual slouches.

And the bones were astonishingly recent! The little arm bone, the first to be excavated, was dated to only 10,000 or 11,000 years ago. The complete skeleton belonged to an adult who had died about 40–50,000 years ago. And the oldest bone fragments, from further down in the excavation, were only

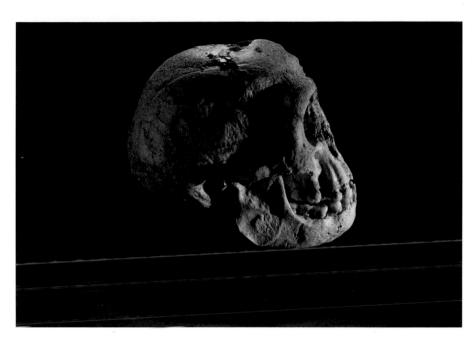

Figure 138 The only complete skull of *Homo floresiensis* yet found shows the small brain case and projecting brow ridges. © Djuna Ivereigh/Arkenas.

about 100,000 years old. The Hobbits must have overlapped with modern humans on Flores.

No other hominan remains have been found from these strata. So the simplest interpretation of these discoveries is that some unknown ancestors of the Hobbits, possibly *H. erectus* but more likely some other hominan, arrived on Flores at least 800,000 years ago. Although they brought with them only primitive tools, they probably improved on them over time. This newly invented technology allowed them to hunt the large animals that they found on the island, some of which were fierce predators. And once they had arrived on the island, unless they were tiny to begin with, they were subject to the same natural selection pressure that produced the tiny elephants on Flores and on many other islands around the world. When food is in short supply large mammals tend to become smaller, and the Hobbits may have been no exception.

The giant Komodo dragons of the island, on the other hand, were selected for greater size. Reptiles are not as limited by food as mammals, because their

Figure 139 Mike Morwood looks into the shaft where the complete Hobbit skeleton was found. Excavations in the cave and at the Soa Basin, along with new excavations on the island of Sulawesi, are likely to cast more light on the origin of the Hobbits and the complicated history of the hominans in the Sunda Islands.

low metabolic rates mean that they can go far longer between meals. And so another type of natural selection took over in the dragons, the process of sexual selection. Larger and fiercer males were selected for because they were more successful at finding and keeping mates. Females too became larger, because the genes for increased size tend to operate in both males and females. Eventually the dragons became big enough to stalk tourists.

The discovery of the Hobbit skeleton was unprecedented. Not only was this tiny female wildly different from any other hominan found outside Africa, but she also had no clear antecedents in the fossil record. Many scientists felt uncomfortable with this huge and unexplained gap in the hominan story.

It only made things worse when, almost instantaneously after the report on the fossils appeared, they were given the name that flashed around the world. The three *Lord of the Rings* movies had just appeared, and it would have been a feeble-minded journalist indeed who failed to make the connection between the tiny woman of Flores and the Hobbits of Middle Earth. The name has stuck and has even worked its way into the scientific literature as a shorthand synonym for the official name, *Homo floresiensis*.

Robert Martin of Chicago's Field Museum posed an alternative hypothesis for the evolution of these little people. He suggested that they were modern humans who had suffered some developmental abnormality such as microcephaly, accompanied by a selection-driven reduction in stature. Dean Falk of Florida State University and her colleagues investigated this possibility. They showed that the Hobbit skulls are clearly different from those of modern-day microcephalics, and their later studies also ruled out other possible pathologies such as Laron Syndrome. As the bones are examined more and more minutely, the list of differences between the skeletons of the Hobbits and those of modern humans continues to grow. The list is now so long that it is clear no single pathology affecting modern humans could have been responsible for all of them.

Figure 140 Digging a second shaft at Liang Bua in search of more bones and artifacts.

The argument continues nonetheless. Personally I find it difficult to imagine that a race of pathologically deformed modern humans survived in a state of nature on Flores for 90,000 years. Not only did they survive, but there is also strong circumstantial evidence that they were able to kill elephants and Komodo dragons and dine extravagantly on them.

The clincher that would decide this argument, of course, would be the deciphering of even a small stretch of Hobbit DNA. The clear genetic difference between modern humans and Neanderthals was first demonstrated on the basis of a bit of Neanderthal mitochondrial DNA only about 300 bases long. If Hobbit DNA sequences should also turn out to be substantially different from those of modern humans, then their origin on another branch of the hominan family tree would be proved. Alternatively, if the sequences were the same as those of modern humans, then the Hobbits would be demoted to mere morphological variants of our own species, like the African pygmies and other small-statured groups of people such as the San.

All attempts at isolating DNA from Hobbit bones have so far failed. Unfortunately, the bones in Liang Bua cave have been so wet for so long that bacterial activity has almost certainly destroyed most traces of their owners' DNA. But new technologies that can sequence single molecules of badly fragmented DNA are now available, and they may yield positive results. And there may be Hobbit bones in other, drier caves, more favorable for DNA preservation, that lie in wait for the paintbrushes and tweezers of the archeologists and the remarkable technologies of the molecular biologists.

Like the pre-Neanderthal hominans that were the first colonists of Europe on the other side of the world, the Hobbits and their ancestors seem to have evolved in isolation for at least a million years. Did they acquire new cultural capabilities during this long period of isolation, despite the fact that there were so few of them? Could they have done so even in the face of calorie limitations that selected them for smaller and smaller bodies and brains?

And could it be that the Hobbits were always small? Were they descendants of the tiny African Australopithecines rather than the much larger *H. erectus*? This might mean that groups of Australopithecines managed to accompany *H. erectus* all the way from Africa. These small-brained hominans would have left that continent 2 million years ago, and walked half a world away.

The more likely possibility is that the *H. erectus* that left Africa were a more diverse collection of hominans than we have supposed. They may have harbored so much genetic variation that it was not difficult for natural selection to produce tiny Hobbits from a mix of *H. erectus* types, once they had reached Flores. There is some evidence for this hypothesis. Finds of *H. erectus* in Georgia in western Asia from 1.8 million years ago have puzzled anthropologists, because it is clear that even though all the skeletons belong to a single group of people, they nonetheless range widely in size.[14] And some of the diverse skulls from Georgia also show characteristics that resemble those of the Australopithecines.

We will not be able to distinguish among these possibilities until we learn something about the genetic history of the Hobbits on Flores and on other islands of the Sunda chain. We also need to find more complete sets of tools and artifacts in the little-explored caves and rock strata of these islands. And there is always the exciting possibility that Hobbits lived in other parts of Southeast Asia, and left additional traces deep beneath the floors of the numerous limestone caves that honeycomb the region.

We know distressingly little about the Neanderthals of Europe and western Asia, despite more than a century and a half of searching and study. If we are lucky, our knowledge of the Hobbits will advance more swiftly. And I can confidently predict that the new discoveries will reinforce Darwin's conclusion that there is nothing unique about human evolution. If modern humans had never appeared, there is no reason to suppose that there could not eventually have been thriving civilizations of Neanderthals or Hobbits.

We will never find out, because these close relatives of ours are gone forever. Sadly, they may have been the last victims of the interspecies warfare that accompanied the spread of modern humans. And yet, the discovery that Neanderthals intermixed with modern humans a hundred thousand years ago (Green et al. *Science* 328:710, 2010), and that some of their genes have lasted down to the present time, holds out the possibility that *H. erectus* and the Hobbits might have bequeathed some of their genes to us as well.

We can get some notion of what the Hobbits might really have been like, and what we may have lost through their disappearance, by observing similar peoples in our own time. The San people of the Kalahari diverged from other modern humans somewhere between 100,000 and 200,000 years ago. Through a combination of luck and determination they survived the onslaught of the Bantu peoples from the north, and the even more murderous attacks of the Boers from across the ocean. The San are clever hunters and observant gatherers. And they are lively and fun-loving, cheerfully relating to each other in ways entirely familiar to any other member of our species.

The San laugh and shout as they gather food in the southern African bush. Did similar laughter and shouts of excitement echo off the walls of the Liang Bua cave? Did the Hobbits laugh with pleasure at some new discovery, and poke cheerful fun at each other? The sound waves generated by those far-off events have long since vanished. But the echo of the Hobbits' essential humanness, and perhaps of a lost evolutionary opportunity, remains.

A Final Postcard

Figure 141 How far we have come on life's journey! For most of the Earth's history, a Darwinian tourist visiting the seashore would have been greeted with a scene like this colony of stromatolites, layered concretions of bacteria and algae that live in the warm salty waters of western Australia's Shark Bay. Compared with those simple days, we live in a time of unparalleled biological complexity and diversity. But if we are to preserve this diversity and survive as a species, we must learn to understand our true place and true role in evolutionary history.

It has been my privilege and pleasure throughout this book to demonstrate how the signs of evolutionary change are plain to see no matter where we look on our planet. Although I have taken you to some exotic places, the influence of evolution can be discovered everywhere. We do not have to travel to the ends of the Earth—even when we stay at home we can be Darwinian tourists. Each field, mountain, reef, and canyon reveals parts of the Earth's long past. The consequences of evolution are all around us and affect our daily lives.

Only through such an evolutionary perspective can we understand the true history of the world and of our species, and how our actions will affect the future of the planet. This understanding is of tremendous importance, because everywhere in the world that our species has gone we have changed the environment and reduced the world's ecological and genetic variation, closing off our options for the future.

At the same time, our own activities have produced evolutionary changes in our own species that we are only just on the verge of comprehending. We must understand the changes that we have brought about in ourselves and in the entire living world if we are to meet the immense challenges of human population growth and environmental degradation in the years and centuries to come. We must see the world through evolutionary eyes. If we let ignorance and the denial of evolution prevail, we are in danger of destroying the living fabric of the world that we treasure and that sustains us.

Notes

Introduction

1 A discussion of the influences that have shaped the Mudumalai forest can be found in Robert John, H. S. Dattaraja, H. S. Suresh, and R. Sukumar, "Density-Dependence in Common Tree Species in a Tropical Dry Forest in Mudumalai, Southern India," *Journal of Vegetation Science* 13 (2002): 45–56.

Chapter 1 Shape-Shifters

1 There are many fabulous images on the web of the creatures that live in the Lembeh Strait. One excellent site has been put together by Doug and Elaine Segar: http://www.reefimages.com/locations/Lembeh.htm

2 Extreme reversals of the usual male and female sexual roles in pipefishes and seahorses are discussed in Adam G. Jones, DeEtte Walker, and John C. Avise, "Genetic Evidence for Extreme Polyandry and Extraordinary Sex-Role Reversal in a Pipefish," *Proceedings of the Royal Society: Biological Sciences* 268 (2001): 2531–5.

3 The common ancestry of echinoderms and chordates is examined in detail in Billie J. Swalla and Andrew B. Smith, "Deciphering Deuterostome Phylogeny: Molecular, Morphological and Palaeontological Perspectives," *Philosophical Transactions of the Royal Society B* 363 (2008): 1557–68.

4 Walter Garstang, a marine biologist at Oxford, wittily summarized his views of the relationships among organisms based on their larval similarities in a book of doggerel, *Larval Forms and Other Zoological Verses* (Oxford: Blackwell, 1951; rpt: Chicago: University of Chicago Press, 1985).

5 The astonishingly complex cascade of events that permits the mantis shrimp to strike so rapidly is investigated in S. N. Patek, B. N. Nowroozi, J. E. Baio, R. L. Caldwell, and A. P. Summers, "Linkage Mechanics and Power Amplification of the Mantis Shrimp's Strike," *Journal of Experimental Biology* 210 (2007): 3677–88.

6 The evidence for the common origin of animal eyes is summarized in a review of a meeting on eye development: Jessica E. Treisman, "How to Make an Eye," *Development* 131 (2004): 3823–7.

7 The protective nature of nudibranchs' bright color results from aposematic coloration, in which the bright colors of prey alert predators to their distastefulness. The signals that predatory fish use, however, are a complex mix, as is shown in studies of fish confronted by bright colors and nudibranch extracts in various combinations. See Raphael Ritson-Williams and Valerie J. Paul, "Marine Benthic Invertebrates Use Multimodal Cues for Defense against Reef Fish," *Marine Ecology Progress Series* 340 (2007): 29–39.

8 Some years ago I discussed the work of UCLA's Harry Jerison and others that showed how the brain sizes of mammals, dinosaurs, and even fish have increased over time because of selective pressures for increasing brainpower. I called my account "Escape From Stupidworld" (*Discover* 14(8) (1993): 54–9), one of my all-time favorite titles.

9 You can see some of the astonishing transformations that the mimic octopus is capable of at various sites on *YouTube*. You will find videos of the octopus changing color and shape dramatically, and imitating a sea snake, a bunch of seaweed, a lionfish, and a flounder.

10 A recent look at some of the difficulties confronting the search for the earliest fossil bacteria is in M. Brasier, N. McLoughlin, O. Green, and D. Wacey, "A Fresh Look at the Fossil Evidence for Early Archaean Cellular Life," *Philosophical Transactions of the Royal Society B* 361 (2006): 887–902.

11 A good survey for the general reader of the search for Precambrian life can be found in Andrew H. Knoll, *Life on a Young Planet: The First Three Billion Years of Evolution on Earth* (Princeton, NJ: Princeton University Press, 2004).

12 The early misclassifications of Burgess shale Cambrian organisms is surveyed by Desmond Collins, "Misadventures in the Burgess Shale," *Nature* 460 (2009): 952–3.

The Wiwaxia radula's affinity to present-day mollusk radulas is examined in A. H. Scheltema, K. Kerth, and A. M. Kuzirian, "Original Molluscan Radula: Comparisons among Aplacophora, Polyplacophora, Gastropoda, and the Cambrian Fossil *Wiwaxia corrugata*," *Journal of Morphology* 257 (2003): 219–45.

13 A thorough examination of the Kimberella fossils, including new information about how they left traces of grazing behind in the rocks, is James G. Gehling, Mary L. Droser, Soren R. Jensen, and Bruce N. Runnegar, "Ediacara Organisms: Relating Form to Function," in Derek E. G. Briggs (ed.), *Evolving Form and Function: Fossils and Development* (New Haven, CT: Yale University Press, 2005), 43–66.

14 Various Precambrian dates based on DNA evidence are compared in Jeffrey S. Levinton, "The Cambrian Explosion: How Do We Use the Evidence?" *Bioscience* 58 (2008): 855–64.

15 Lindquist's study of *Drosophila* development and chaperonins is S. L. Rutherford and S. Lindquist, "Hsp90 as a Capacitor for Morphological Evolution," *Nature* 396 (1998): 336–42.

16 Geoffroy Saint-Hilaire's attempts to find a common plan for all animals are traced in Hervé Le Guyader's *Geoffroy Saint-Hilaire, A Visionary Naturalist* (Chicago: University of Chicago Press, 2004). Saint-Hilaire battled Georges Cuvier, who thought that different groups of animals had been created separately and were unrelated to each other.

17 The evolving computer life forms of Lipson and Pollack are detailed in Hod Lipson and Jordan B. Pollack, "Automatic Design and Manufacture of Robotic Lifeforms," *Nature* 406 (2000): 974–8.

2 The Inner Workings of Evolution

1 The discovery of melanin differences in fungi living on the two sides of Evolution Canyon is detailed in Natarajan Singaravelan, Isabella Grishkan, Alex Beharav, Kazumasa Wakamatsu, Shosuke Ito, and Eviatar Nevo, "Adaptive Melanin Response of the Soil Fungus *Aspergillus niger* to UV Radiation Stress at 'Evolution Canyon', Mount Carmel, Israel," *PLoS One* 3 (2008): e2993 (published online).

2 Korol's study of mating preferences in Drosophila on both sides of the canyon is Shree Ram Singh, Eugenia Rashkovetsky, Konstantin Iliadi, Eviatar Nevo, and Abraham Korol, "Assortative Mating in Drosophila Adapted to a Microsite Ecological Gradient," *Behavior Genetics* 35 (2005): 753–64.

3 Korol and Nevo's study of differences in mutation repair in these flies and their correlation with the severity of the environment is Achsa Lupu, Antonina Pechkovskaya, Eugenia Rashkovetsky, Eviatar Nevo, and Abraham Korol, "DNA Repair Efficiency and Thermotolerance in *Drosophila melanogaster* from 'Evolution Canyon'," *Mutagenesis* 19 (2004): 383–90.

4 The association between chromosome number and ecology in different species of naked mole rats is detailed in R. Arieli, M. Arieli, G. Heth, and E. Nevo, "Adaptive Respiratory Variation in Four Chromosomal Species of Mole Rats," *Experientia* 40 (1984): 512–14.

5 The problems with figuring out the closest relatives to the hoatzin are set out in Michael D. Sorenson, Elen Oneal, Jaime Garcia-Moreno, and David P. Mindell, "More Taxa, More Characters: The Hoatzin Problem Is Still Unresolved," *Molecular Biology and Evolution* 20 (2003): 1484–98.

6 The hoatzin crop is home to an astonishing 1800 different species of bacteria. Few of then are shared with other fermenting organisms such as the ruminants. Details of this remarkable community are found in Filipa Godoy-Vitorino, Ruth E. Ley, Zhan Gao, Zhiheng Pei, Ortiz- Humberto Zuazaga, et al.,

"Bacterial Community in the Crop of the Hoatzin, a Neotropical Folivorous Flying Bird," *Applied and Environmental Microbiology* 74 (2008): 59–5912.

7 The detailed feeding habits of a troop of proboscis monkeys are examined in Ikki Matsuda, Augustine Tuuga, and Seigo Higashi, "The Feeding Ecology and Activity Budget of Proboscis Monkeys," *American Journal of Primatology* 71 (2009): 478–92.

8 The catholic feeding habits of capuchin monkeys are investigated in Carlos Henrique de Freitas, Eleonore Z. F Setz, Alba R. B. Araujo, et al., "Agricultural Crops in the Diet of Bearded Capuchin Monkeys, *Cebus libidinosus Spix* (Primates: Cebidae), in Forest Fragments in Southeast Brazil," *Revista Brasileira de Zoologia* 25 (2008): 32–9.

9 Allan Wilson's first investigations of the evolutionary relationships among lysozymes are summarized in D, M. Irwin, E. M. Prager, and A. C. Wilson, "Evolutionary Genetics of Ruminant Lysozymes," *Animal Genetics* 23 (1992): 193–202.

10 Pathbreaking comparisons of lysozymes in different groups of animals were carried out by Caro-Beth Stewart, James W. Schilling, and Allan C. Wilson, "Adaptive Evolution in the Stomach Lysozymes of Foregut Fermenters," *Nature* 330 (1987): 401–4.

11 The convergent evolution of hoatzin and mammal lysozymes is explored in Janet R. Kornegay, James W. Schilling, and Allan C. Wilson, "Molecular Adaptation of a Leaf-Eating Bird: Stomach Lysozyme of the Hoatzin," *Molecular Biology and Evolution* 11 (1994): 921–8.

3 The Shifting Earth

1 The remarkable story of the Yapese people and their stone money is recounted in *His Majesty O'Keefe*, by Lawrence Klingman and Gerald Green (Scribner, 1950), a novelized version of the true story of nineteenth-century Irish-American adventurer David O'Keefe, who persuaded the people of the island to work for him by rewarding them with new supplies of the stone money from the island of Babeldaub.

2 I recounted the story of the Yap earthquake in the April–June 2008 issue of *Asia-Pacific Alert Diver*, the magazine of the Alert Divers Network, pp. 18–19.

3 Darwin told of his earthquake experience in *The Voyage of the Beagle*, his fascinating account of the journey. First published as Volume 3 of *Narrative of the Surveying Voyages of His Majesty's Ships Adventure and Beagle* in 1839, it has been reprinted numerous times as a single volume since.

4 A cautionary passage from Darwin's *Autobiography* (published after his death in 1887):

On this tour I had a striking instance of how easy it is to overlook phenomena, however conspicuous, before they have been observed by any one. We spent many hours in Cwm Idwal, examining all the rocks with extreme care, as Sedgwick was anxious to find fossils in them; but neither of us saw a trace of the wonderful glacial phenomena all around us; we did not notice the plainly scored rocks, the perched boulders, the lateral and terminal moraines. Yet these phenomena are so conspicuous that, as I declared in a paper published many years afterwards in the "Philosophical Magazine" [Philosophical Magazine, 1842], a house burnt down by fire did not tell its story more plainly than did this valley. If it had still been filled by a glacier, the phenomena would have been less distinct than they now are.

5 The discovery of Antarctic *Glossopteris* is detailed in Campbell Craddock, Thomas W. Bastien, Robert H. Rutford, and John J. Anderson, "Glossopteris Discovered in West Antarctica," *Science* 148 (1965): 634–7. The Antarctic *Glossopteris* discovery was predicted by A. L. du Toit in *Our Wandering Continents* (London: Bradford and Dickens, 1937).

6 Arthur Holmes read a paper in 1927 to the Edinburgh Geological Society in which he suggested that the Earth's internal radioactivity was sufficient to power great convective cells of molten magma that could drive the continents apart. His ideas were not fully accepted until the 1960s, and they have been substantially modified since that time.

7 The story of early mammal evolution is detailed in Zhe-Xi Luo, "Transformation and Diversification in Early Mammal Evolution," *Nature* 450 (2007): 1011–19. The author points out that the fossils of many genera of mammals that lived during the Age of Dinosaurs have been discovered—almost half as many as the genera of dinosaurs themselves!

   The long "fuse" that started mammalian radiation and that leads back to before the beginning of the Age of Mammals 65 million years ago is discussed in Olaf R. P. Bininda-Emonds, Marcel Cardillo, Kate E. Jones, Ross D. E. MacPhee, Robin M. D. Beck, Richard Grenyer, *et al.*, "The Delayed Rise of Present-Day Mammals," *Nature* 446 (2007): 507–12. It is striking that the pattern of early mammalian radiation is reminiscent of the radiation of animals before the beginning of the Cambrian, except that we have both DNA and fossil evidence for the more recent origin of mammals.

8 A somewhat dated but fascinating exploration of the invasion of South America by placental animals can be found in George Gaylord Simpson, *Splendid Isolation: The Curious History of South American Mammals* (New Haven, CT: Yale University Press, 1983).

9 Stanley Ambrose's original suggestion about the deadly effects of the Toba eruption is set out in Stanley H. Ambrose, "Late Pleistocene Human Population Bottlenecks, Volcanic Winter, and Differentiation of Modern Humans," *Journal of Human Evolution* 34 (1998): 623–51.

10 Michel Petraglia's observations on the continuity of Middle Paleolithic tools in India before and after the eruption are in Michael Petraglia, Ravi Korisettar, Nicole Boivin, Christopher Clarkson, Peter Ditchfield, *et al.*, "Middle Paleolithic Assemblages from the Indian Subcontinent Before and After the Toba Super-Eruption," *Science* 317 (2007): 114–16.

11 The first paper noting the increasing diversity over time during the geological record is J. John Sepkoski, Richard K. Bambach, David M. Raupand, and James W. Valentine, "Phanerozoic Marine Diversity and the Fossil Record," *Nature* 293 (1981): 435–7. A correlation of this diversity with increased signs of predation was found by Steven M. Holland, "Coupling of Predation Intensity and Global Diversity over Geologic Time," *Proceedings of the National Academy of Sciences (U.S.)* 104 (2007): 14885–6. This relationship strongly suggests that an increase in the number and complexity of ecological niches, not necessarily entirely due to predation but nonetheless correlated with increased predation, has contributed to the dramatic increase in diversity that is seen in the fossil record.

4 Crucibles of Speciation

1 The story of the finding of a new lemur and the rediscovery of an old one is told in Patricia Wright, "Lemurs Lost and Found," *Natural History* (July 1988), 56–60.

2 The story of how the different amounts of cyanide in bamboos has dictated the diets of the bamboo lemurs is detailed in K. E. Glander, P. C. Wright, D. S. Seigler, V. Randrianasolo, and B Randrianasolo, "Consumption of Cyanogenic Bamboo by a Newly Discovered Species of Bamboo Lemur," *American Journal of Primatology* 19 (1989): 119–124, and Daniel J. Ballhorn, Stefanie Kautz, and Fanny P. Rakotoarivelo, "Quantitative Variability of Cyanogenesis in *Cathariostachys madagascariensis*—The Main Food Plant of Bamboo Lemurs in Southeastern Madagascar," *American Journal of Primatology* 71 (2009): 305–15.

3 Dung beetles and lemurs in Madagascar have speciated together, as shown in Helena Wirta, Luisa Orsini, and Ilkka Hanski, "An Old Adaptive Radiation of Forest Dung Beetles in Madagascar," *Molecular Phylogenetics and Evolution* 47 (2008): 1076–89.

4 The sensitivity of Madagascar beetle species to deforestation is detailed in Ilkka Hanski, Helena Koivulehto, Alison Cameron, and Pierre Rahagalala, "Deforestation and Apparent Extinctions of Endemic Forest Beetles in Madagascar," *Biology Letters* 3 (2007): 344–7.

5 Neil Rooney sets out a frequency-dependent model for the role of top predators in controlling numbers in complex ecological food webs in Neil Rooney, Kevin McCann, Gabriel Gellner, and John C. Moore, "Structural Asymmetry and the Stability of Diverse Food Webs," *Nature* 442 (2006): 265–9.

6 The remarkable studies of the ecology and genetics of Darwin's finches in the Galàpagos Islands that have been carried out by Peter and Rosemary Grant and their many colleagues are detailed in Peter R. Grant and B. Rosemary Grant, *How and Why Species Multiply: The Radiation of Darwin's Finches* (Princeton, NJ: Princeton University Press, 2007).

5 Rainforests, Diseases, and Speciation

1 A. R. Wallace remarked, in *Tropical Nature* (Macmillan, 1878):

> *The primeval forests of the equatorial zone are grand and overwhelming by their vastness, and by the display of a force of development and vigour of growth rarely or never witnessed in temperate climates. Among their best distinguishing features are the variety of forms and species which everywhere meet and grow side by side, and the extent to which parasites, epiphytes and creepers fill up every available station with peculiar modes of life. If the traveler notices a particular species and wishes to find more like it, he may often turn his eyes in vain in every direction. Trees of varied forms, dimensions, and colours are around him, but he rarely sees any one of them repeated. Time after time he goes towards a tree which looks like the one he seeks, but a closer examination proves it to be distinct. He may at length, perhaps, meet with a second specimen half a mile off, or may fail altogether, till on another occasion he stumbles on one by accident.*

2 The early Colorado rainforest was described in Kirk R. Johnson and Beth Ellis, "A Tropical Rainforest in Colorado 1.4 Million Years after the Cretaceous-Tertiary Boundary," *Science* 296 (2002): 2379–86.

3 Bruce Tiffney's discovery of the sudden increase in angiosperm seed sizes is in Bruce H. Tiffney, "Seed Size, Dispersal Syndromes, and the Rise of the Angiosperms: Evidence and Hypothesis," *Annals of the Missouri Botanical Garden* 71 (1984): 551–76.

4 The seminal papers by Janzen and Connell on the maintenance of rainforest diversity by predators and pathogens are D. H. Janzen, "Herbivores and the Number of Tree Species in Tropical Forests," *American Naturalist* 104 (1970), 501–28, and J. H. Connell, "On the Role of Natural Enemies in Preventing Competitive Exclusion in Some Marine Animals and in Rain Forest Trees," in P. J. Den Boer and G. Gradwell (eds), *Dynamics of Populations* (Pudoc (Wageningen, Netherlands), 1971), 298–312.

5 The elegant experiments showing the Janzen-Connell effect in black cherry trees are Alyssa Packer and Keith Clay, "Soil Pathogens and Spatial Patterns of Seedling Mortality in a Temperate Tree," *Nature* 404 (2000): 278–81.

6 The basic idea of niche complementarity was introduced in Earl E. Werner, "Species Packing and Niche Complementarity in Three Sunfishes," *American Naturalist* 111 (1977): 553–78.

7 The history of the Smithsonian forest plots, starting with the Barro Colorado Island plot, is entertainingly set forth in a piece by Wendy Madar of Oregon State University, to be found at http://www.sigeo.si.edu/data///documents/Madar_CTFS_Way.pdf

8 My analysis of the Barro Colorado Island plot data is in Christopher Wills, Richard Condit, Robin B. Foster, and Stephen P. Hubbell, "Strong Density- and Diversity-Related Effects Help to Maintain Tree Species Diversity in a Neotropical Forest," *Proceedings of the National Academy of Sciences (U.S.)* 94 (1997): 1252–7.

9 Our survey that shows the active maintenance of diversity in tropical forests around the world is Christopher Wills, Kyle E. Harms, Richard Condit, David King, Jill Thompson, Fangliang He, Helene C. Muller-Landau, Peter Ashton, Elizabeth Losos, *et al.*, "Nonrandom Processes Maintain Diversity in Tropical Forests," *Science* 311 (2006): 527–31.

10 Some of the remarkable work of Doug Schenske and his colleagues on the genetics of monkey-flower speciation can be found in H. D. Bradshaw and D. W. Schemske, "Allele Substitution at a Flower Colour Locus Produces a Pollinator Shift in Monkeyflowers," *Nature* 426 (2003): 176–8.

11 The work of Ajit Varki and his colleagues on the evolution of the sugar molecules that stud our cells is reviewed in Ajit Varki, "Multiple Changes in Sialic Acid Biology during Human Evolution," *Glycoconjugate Journal* 26 (2009): 231–45.

12 The origin of *Plasmodium falciparum* is explored in Stephen M. Rich, Fabian H. Leendertz, Guang Xu, Matthew LeBreton, *et al.*, "The Origin of Malignant Malaria," *Proceedings of the National Academy of Sciences (U.S.)* 106 (2009): 14902–7. The connection of this work to sialic acid metabolism is suggested by Ajit Varki and Pascal Gagneux, "Human-Specific Evolution of Sialic Acid Targets: Explaining the Malignant Malaria Mystery?" *Proceedings of the National Academy of Sciences (U.S.)* 106 (2009): 14739–40.

6 How Domesticated Animals Changed the World

1 The differences in how dogs and wolves respond to humans is explored in Brian Hare, Michelle Brown, Christina Williamson, and Michael Tomasello, "The Domestication of Social Cognition in Dogs," *Science* 298 (2002): 1634–6.

2 Peter Savolainen's investigation into dog origins is Peter Savolainen, Ya-ping Zhang, Jing Luo, Joakim Lundeberg, and Thomas Leitner, "Genetic Evidence for an East Asian Origin of Domestic Dogs," *Science* 298 (2002): 1610–13.

3 The careful redating of a remarkably old Belgian dog skull is recounted in Mietje Germonpre, Mikhail V. Sablin, Rhiannon E. Stevens, Robert E. M. Hedges, *et al.*, "Fossil Dogs and Wolves from Palaeolithic Sites in Belgium, the Ukraine, and Russia: Osteometry, Ancient DNA and Stable Isotopes," *Journal of Archaeological Science* 36 (2009): 473–90.

4 Investigations of dog DNA from Precolumbian New World burial sites have shown that dogs accompanied the first humans to the Americas: Jennifer A. Leonard, Robert K. Wayne, Jane Wheeler, Raúl Valadez, Sonia Guillén, and Carles Vilà, "Ancient DNA Evidence for Old World Origin of New World Dogs," *Science* 298 (2002): 1613–16.

5 The variety of uses of dingoes by Aboriginal Australians is discussed in Tim Flannery, *The Future Eaters: An Ecological History of the Australasian Lands and People* (New York: Grove Press, 1994).

6 Bustamante's study on the high diversity in African village dogs is Adam R. Boykoa, Ryan H. Boykob, Corin M. Boykob, Heidi G. Parker, Marta Castelhanod, *et al.*, "Complex Population Structure in African Village Dogs and Its Implications for Inferring Dog Domestication History," *Proceedings of the National Academy of Sciences (U.S.)* 106 (2009): 13903–8.

7 Paul Martin's theory proposing many human-caused extinctions is Paul Martin, "Pleistocene Overkill," *Natural History* 76 (1967): 32–8. His reply to criticisms of his theory that center on the apparent absence of African extinctions is Paul Martin, "Africa and Pleistocene Overkill," *Nature* 212 (1966): 339–42.

8 The timing of elephant overkill throughout the temperate zones strongly suggests human involvement. The evidence for this is presented in Todd Surovell, Nicole Waguespack, and P. Jeffrey Brantingham, "Global Archaeological Evidence for Proboscidean Overkill," *Proceedings of the National Academy of Sciences (U.S.)* 102 (2005): 6231–6.

9 A detailed look at the Maori dogs of New Zealand can be found in Geoffrey Clark, "Osteology of the Kuri Maori: The Prehistoric Dog of New Zealand," *Journal of Archaeological Science* 24 (1997): 113–26.

10 The separate history of Indian wolves, and their distinction from Indian dogs, is investigated in Dinesh K. Sharma, Jesus E. Maldonado, Yadrendradev V. Jhala, and Robert C. Fleischer, "Ancient Wolf Lineages in India," *Proceedings of the Royal Society B (Suppl.)* 271 (2004): S1–S4.

11 The genetic basis of size differences among dog breeds is explained by Nathan B. Sutter, Carlos D. Bustamante, Kevin Chase, Melissa M. Gray, *et al.*, "A Single IGF1 Allele is a Major Determinant of Small Size in Dogs," *Science* 316 (2007): 112–15.

12 The adaptation of yaks to high altitude is detailed in Anthony G. Durmowicz, Stephen Hofmeister, T. K. Kadyraliev, *et al.*, "Functional and Structural Adaptation of the Yak Pulmonary Circulation to Residence at High Altitude," *Journal of Applied Physiology* 74 (1993): 2276–85.

13 The similar adaptation of Tibetan people to high altitudes, and its effect on birth weight of infants, is investigated in Lorna G. Moore, David Young, Robert E. McCullough, *et al.*, "Tibetan Protection from Intrauterine Growth Restriction (IUGR) and Reproductive Loss at High Altitude," *American Journal of Human Biology* 13 (2001): 635–44.

14 The genetic history of yaks is traced in Songchang Guo, Peter Savolainen, Jianping Su, Qian Zhang, *et al.*, "Origin of Mitochondrial DNA Diversity of Domestic Yaks," *BMC Evolutionary Biology* 6 (2006): 73–86.

15 The history of aurochs and cattle is examined in Ceiridwen J. Edwards, Ruth Bollongino, Amelie Scheu, Andrew Chamberlain, *et al.*, "Mitochondrial DNA Analysis Shows a Near Eastern Neolithic Origin for Domestic Cattle and No Indication of Domestication of European Aurochs," *Proceedings of the Royal Society B* 274 (2007): 1377–85.

16 Caesar, *The Gallic Wars*:
    There is a third kind, consisting of those animals which are called uri. These are a little below the elephant in size, and of the appearance, colour, and shape of a bull. Their strength and speed are extraordinary; they spare neither man nor wild beast which they have espied. These the Germans take with much pains in pits and kill them. The young men harden themselves with this exercise, and practice themselves in this kind of hunting, and those who have slain the greatest number of them, having produced the horns in public, to serve as evidence, receive great praise. But not even when taken very young can they be rendered familiar to men and tamed. The size, shape, and appearance of their horns differ much from the horns of our oxen. These they anxiously seek after, and bind at the tips with silver, and use as cups at their most sumptuous entertainments.

17 An account of the grim ecological effect of the zebu in Madagascar is found in D. W. Gade, "Deforestation and Its Effects in Highland Madagascar," *Mountain Research and Development* 16 (1996): 101–16.

18 The domestication of goats and sheep is traced in Gordon Luikart, Ludovic Gielly, Laurent Excoffier, Jean-Denis Vigne, *et al.*, "Multiple Maternal Origins and Weak Phylogeographic Structure in Domestic Goats," *Proceedings of the National Academy of Science (U.S.)* 98 (2001): 5927–32, and Jennifer R. S. Meadows, Ibrahim Cemal, Orhan Karaca, Elisha Gootwine, and James W. Kijas, "Five Ovine Mitochondrial Lineages Identified from Sheep Breeds of the Near East," *Genetics* 175 (2007): 1371–9.

19 Baker and Manwell's theory of partial elephant domestication and how it might have happened can be found in C. M. Ann Baker and Clyde Manwell, "Man and Elephant: The 'Dare Theory' of Domestication and the Origin of Breeds," *Zeitschrift für Tierzüchtungs und Züchtungsbiologie* 100 (1983): 55–75.

20 Gavin de Beer's examination of the origin of Hannibal's elephants is in Gavin de Beer, *Alps and Elephants: Hannibal's March* (London: Geoffrey Bles, 1955).

21 As we saw earlier, the dating of butchering sites from the temperate zone Paleolithic suggests human involvement in the recent extinctions of elephants and their relatives. The evidence is presented in Todd Surovell, Nicole Waguespack, and P. Jeffrey Brantingham, "Global Archaeological Evidence for Proboscidean Overkill," *Proceedings of the National Academy of Sciences (U.S.)* 102 (2005): 6231–6.

22 A recent census of elephants at Garamba can be found in Emmanuel de Merode, Bila-Isia Inogwabini, José Telo, and Ginengayo Panziama, "Status of Elephant Populations in Garamba National Park, Democratic Republic of Congo, late 2005," *Pachyderm* Issue 42 (2007): 52–7.

23 The complex and intertwined history of domestic and Przewalski's horses is examined in Allison N. Lau, Lei Peng, Hiroki Goto, Leona Chemnick, Oliver A. Ryder, and Kateryna D. Makova, "Horse Domestication and Conservation

Genetics of Przewalski's Horse Inferred from Sex Chromosomal and Autosomal Sequences," *Molecular Biology and Evolution* 26 (2009): 199–208.

24 The origins of the Bactrian two-humped camels are traced in R. Ji, P. Cui, F. Ding, J. Geng, *et al.*, "Monophyletic Origin of Domestic Bactrian camel (*Camelus bactrianus*) and Its Evolutionary Relationship with the Extant Wild Camel (*Camelus bactrianus ferus*)," *Animal Genetics* 40 (2009): 377–82.

25 In the nineteenth and early twentieth centuries zebras were harnessed to carriages in both Africa and Europe. But they had to be blinkered and carefully confined between traces. Nobody rode zebras, not so much because they were vicious as because they were subject to unexpected fits of fear. At the turn of the twentieth century, James Cossar Ewart hybridized horses and zebras to produce tractable and useful animals, but the male hybrids were sterile and the females reproduced poorly.

26 The story of the unusually common Mongolian Y chromosome and its genetic implications is told in Tatiana Zerjal, Yali Xue, Giorgio Bertorelle, R. Spencer Wells, *et al.*, "The Genetic Legacy of the Mongols," *American Journal of Human Genetics* 72 (2003): 717–21.

27 One of a growing number of pieces of evidence showing that our evolution has accelerated is John Hawks, Eric T. Wang, Gregory M. Cochran, Henry C. Harpending, and Robert K. Moyzis, "Recent Acceleration of Human Adaptive Evolution," *Proceedings of the National Academy of Sciences (U.S.)* 104 (2007): 20753–8. Let me be uncharacteristically immodest for a moment, and point out that I anticipated these recent discoveries about the increasing rate of human evolution a decade ago: Christopher Wills, *Children of Prometheus: the Accelerating Pace of Human Evolution* (Cambridge, MA: Perseus Books, 1998).

7 The Great Migration

1 The uncertainties of estimating the earliest human occupation of Australia's Northern Territory are set out in: J. F. O'Connell and J. Allen, "Dating the Colonization of Sahul (Pleistocene Australia–New Guinea): A Review of Recent Research," *Journal of Archaeological Science* 31 (2004), 835–53.

2 A tracing of the Great Migration, based on complete mitochondrial sequences from a number of populations along the route, is given in Stephen Oppenheimer, "The Great Arc of Dispersal of Modern Humans: Africa to Australia," *Quaternary International* 202 (2009): 2–13.

3 The city of Darwin was named after Charles Darwin in 1839 by John Wickham, who was then the commander of the *Beagle*. Wickham had been Lieutenant on the *Beagle*, under Robert FitzRoy, during the little ship's earlier voyage with Darwin. It was on the *Beagle*'s next voyage, long before Darwin became famous, that Wickham named what was then a tiny watering place after his friend.

4 Rock art showing human and animal figures in northern Australia goes back at least 17,000 years, as revealed by thermoluminescence dating of sand grains in nests that wasps built on the surface of old paintings! See Richard Roberts, Grahame Walsh, Andrew Murray, Jon Olley, *et al.*, "Luminescence Dating of Rock Art and Past Environments Using Mud-Wasp Nests in Northern Australia," *Nature* 387 (1997): 696–9.

5 An axe head with associated bitumen, almost certainly Neanderthal in origin, was found at a site in Syria: Eric Boëda, Jacques Connan, Daniel Dessort, Sultan Muhesen, *et al.*, "Bitumen as a Hafting Material on Middle Palaeolithic Artifacts," *Nature* 380 (1996): 336–8.

6 The discovery of a burial involving flowers at the Shanidar site in Iraq was announced in Arlette Leroi-Gourhan, "The Flowers Found with Shanidar IV, a Neanderthal Burial in Iraq," *Science* 190 (1975): 562–4. The flowers, of seven different species, must have been brought into the cave at the time of the burial, which was positioned 15 meters inside the cave entrance.

7 The extent of genetic mixing between the northern and southern populations of India is examined in David Reich, Kumarasamy Thangaraj, Nick Patterson, Alkes L. Price, and Lalji Singh, "Reconstructing Indian Population History," *Nature* 461 (2009): 489–94. It is clear from this study that Indian tribal populations have retained a smaller contribution from the peoples of the Great Migration than the Andaman Islanders, who have been isolated for longer.

8 The mitochondrial DNA study that clarifies both the timing and the sequence of the events of the Great Migration is Vincent Macaulay, Catherine Hill, Alessandro Achilli, Chiara Rengo, *et al.*, "Single, Rapid Coastal Settlement of Asia Revealed by Analysis of Complete Mitochondrial Genomes," *Science* 308 (2005): 1034–6.

9 The remarkably recent dates for *Homo erectus* on Java were found by C. C. Swisher III, W. J. Rink, S. C. Antón, H. P. Schwarcz, *et al.*, "Latest *Homo erectus* of Java: Potential Contemporaneity with *Homo sapiens* in Southeast Asia," *Science* 274 (1996): 870–4.

10 There is some evidence from Java's cave of Song Gupuh that humans arrived early on that island, but it is circumstantial: M. J. Morwood, T. Sutikna, E. W. Saptorno, and K. E. Westaway, "Climate, People and Faunal Succession on Java, Indonesia: Evidence from Song Gupuh," *Journal of Archaeological Science* 35 (2008), 1776–89. The first signs of modern human technology and trading between the coast and the interior of the island appear 12,000 years ago.

11 The work of Graeme Barker and his colleagues at Niah Caves is summarized in G. Barker, P. J. Piper, and R. J. Rabett, "Zooarchaeology at the Niah Caves, Sarawak: Context and Research Issues," *International Journal of Osteoarchaeology* 19 (2009): 447–63.

12 The story of how the Murnong plant was harvested and used for food is detailed in Beth Gott, "Indigenous Use of Plants in South-Eastern Australia," *Telopea* 12 (2008): 215–26.

13 Some of the finds of Middle East stone tools, at sites inland from the probable path of the Great Migration, are detailed in Robin W. Dennell, Helen M. Rendell, Mohammad Halim, and Eddie Moth, "A 45,000-Year-Old Open-Air Paleolithic Site at Riwat, Northern Pakistan," *Journal of Field Archaeology* 19 (1992): 17–33. The Semnan finds from Iran are recent and have not yet been formally published.

14 The discovery that hippos survived until recent times on the west coast of India is reported in: R. W. Dennell, "Early Pleistocene Hippopotamid Extinctions, Monsoonal Climates, and River System Histories in South and South West Asia: Comment on Jablonski (2004), 'The Hippo's Tale: How the Anatomy and Physiology of Late Neogene Hexaprotodon Shed Light on Late Neogene Environmental Change,'" *Quaternary International* 126–8 (2005): 283–7.

8 The San and the Hobbits

1 Van der Post's book is *The Lost World of the Kalahari* (New York: Morrow, 1958).

2 A human mitochondria DNA tree, based on 52 complete mitochondrial sequences, unequivocally shows the origin of our species in Africa: G. Barker, D. Badang, H. Barton, P. Beavitt, *et al.*, "Mitochondrial Genome Variation and the Origin of Modern Humans," *Nature* 408 (2000): 708–12.

3 The discovery of the earliest currently known modern humans from Africa is recounted in Tim D. White, Berhane Asfaw, David DeGusta, Henry Gilbert, *et al.*, "Pleistocene *Homo sapiens* from Middle Awash, Ethiopia," *Nature* 423 (2003): 742–7. These early people were part of a continuum of human evolution, and certainly not identical to us.

4 The suspiciously old San paintings in the Apollo 11 cave in Namibia are discussed in John Masson, "Apollo 11 Cave in Southwest Namibia: Some Observations on the Site and Its Rock Art," *South African Archaeological Bulletin* 61 (2006): 76–89.

5 The revisionist reanalysis that reveals the long history of human culture in Africa is S. McBrearty and A. S. Brooks, "The Revolution That Wasn't: A New Interpretation of the Origin of Modern Human Behavior," *Journal of Human Evolution* 39 (2000): 453–563.

6 The oldest European hominan is reported in: Eudald Carbonell, José M. Bermúdez de Castro, Josep M. Parés, Alfredo Pérez-González, et al., "The First Hominin of Europe," *Nature* 452 (2008): 465–9. Note that "hominin" is a taxonomic category called a tribe that includes the human lineage and chimpanzees, while the narrower subtribe "hominan" used in this book excludes chimpanzees but does include all our close extinct relatives.

7 The complex nature of human–chimpanzee speciation, and the possibility that it took place over substantial periods of time with exchanges of genes during the process, emerged from comparisons of the human and chimpanzee genomes. The hypothesis is presented in Nick Patterson, Daniel J. Richter, Sante Gnerre, Eric S. Lander, and David Reich, "Genetic Evidence for Complex Speciation of Humans and Chimpanzees," *Nature* 441 (2006), 1103–08.

8 The discovery of pre-Neanderthal spears is documented in H. Thieme, "Lower Palaeolithic Hunting Spears from Germany," *Nature* 385 (1997): 807–10.

9 Wallace's rather embarrassing musings about human spirituality can be found in Alfred Russel Wallace, *Contributions to the Theory of Natural Selection: A Series of Essays* (New York: Macmillan, 1871), 372.

10 The successful dating of the stone tools at Mata Menge is M. J. Morwood, P. B. O'Sullivan, F. Aziz, and A. Raza, "Fission-Track Ages of Stone Tools and Fossils on the East Indonesian Island of Flores," *Nature* 392 (1998): 173–6.

11 The discovery of the complete Hobbit skeleton was announced in P. Brown, T. Sutikna, M. J. Morwood, R. P. Soejono, *et al.*, "A New Small-Bodied Hominin from the Late Pleistocene of Flores, Indonesia," *Nature* 431 (2004): 1055–61.

12 Matt Tocheri's analysis of the Hobbits' non-human wrist bones is Matthew W. Tocheri, Caley M. Orr, Susan G. Larson, *et al.*, "The Primitive Wrist of *Homo floresiensis* and Its Implications for Hominin Evolution," *Science* 317 (2007): 1743–5.

13 The first comparison of the Hobbit endocast with pathological modern human brains can be found in D. Falk, C. Hildebolt, K. Smith, M.J. Morwood, *et al.*, "The Brain of LB1, *Homo floresiensis*," *Science* 308 (2005): 242–5.

14 A recent examination of the Georgia *Homo erectus* finds, showing remarkable morphological variation among individuals, is presented in Maria Martinon-Torres, Jose Maria Bermudez de Castro, Aida Gomez-Robles, Ann Margvelash-vili, *et al.*, "Dental Remains from Dmanisi (Republic of Georgia): Morphological Analysis and Comparative Study," *Journal of Human Evolution* 55 (2008): 249–73.

Index